Intelligent Systems Reference Library 41

For further volumes:
http://www.springer.com/series/8578

Igor Chikalov, Vadim Lozin, Irina Lozina,
Mikhail Moshkov, Hung Son Nguyen,
Andrzej Skowron, and Beata Zielosko

Three Approaches to Data Analysis

Test Theory, Rough Sets and Logical Analysis of Data

Authors
Igor Chikalov
Mathematical and Computer Sciences and Engineering Division
King Abdullah University of Science and Technology
Thuwal
Saudi Arabia

Vadim Lozin
Mathematics Institute
The University of Warwick
Coventry
United Kingdom

Irina Lozina
Mathematics Institute
The University of Warwick
Coventry
United Kingdom

Mikhail Moshkov
Mathematical and Computer Sciences and Engineering Division
King Abdullah University of Science and Technology
Thuwal
Saudi Arabia

Hung Son Nguyen
Institute of Mathematics
The University of Warsaw
Warsaw
Poland

Andrzej Skowron
Institute of Mathematics
The University of Warsaw
Warsaw
Poland

Beata Zielosko
Mathematical and Computer Sciences and Engineering Division
King Abdullah University of Science and Technology
Thuwal
Saudi Arabia

and

Institute of Computer Science
University of Silesia
Sosnowiec
Poland

ISSN 1868-4394
e-ISSN 1868-4408
ISBN 978-3-642-44598-9
ISBN 978-3-642-28667-4 (eBook)
DOI 10.1007/978-3-642-28667-4
Springer Heidelberg New York Dordrecht London

Printed on acid-free paper

Springer is part of Springer Science+Business Media (www.springer.com)

To the memory of

Peter L. Hammer,

Zdzisław I. Pawlak

and

Sergei V. Yablonskii

Preface

In this book, we consider the following three approaches to data analysis:

- Test Theory (TT), founded by Sergei V. Yablonskii (1924-1998); the first publications [5,11] appeared in 1955 and 1958,
- Rough Sets (RS), founded by Zdzisław I. Pawlak (1926-2006); the first publications [8,9] appeared in 1981 and 1982 (see, also, the book by Pawlak [10]),
- Logical Analysis of Data (LAD), founded by Peter L. Hammer (1936-2006); the first publications [6,7] appeared in 1986 and 1988.

These three approaches have much in common. For example, they are all related to Boolean functions and Boolean reasoning with the roots in works by George Boole (see, *e.g.*, [1-4]). However, we quite often observe that researchers active in one of these areas have a limited knowledge about the results and methods developed in the other two. On the other hand, each of the approaches shows some originality and we believe that the exchange of knowledge can stimulate further development of each of them. This can lead to new theoretical results and real-life applications. In particular, we expect new results based on combination of these three data analysis approaches.

It would be very interesting to make a comprehensive comparative analysis of the three approaches. However, in the present book, we restrict ourselves to a simpler task and present only a detailed overview of each of the three approaches. To make the reading easier, in the preface, we give a brief comparison of the main notions used in TT, RS and LAD, and a short outline of the overviews.

All three data analysis approaches use *decision tables* for data representation. A decision table T is a rectangular table with n columns labeled with conditional attributes $a_1,\ldots,a_n$. This table is filled with values of (conditional) attributes $a_1,\ldots,a_n$, and each row of the table is labeled by a value of the decision attribute d. There are different types of data and different problems associated with them.

A decision table is said to be *consistent*, if each combination of values of conditional attributes uniquely determines the value of the decision attribute, and inconsistent, otherwise. In TT and LAD, only consistent decision tables are considered, while RS allows inconsistent decision tables. In TT, decision tables are also named *test tables*.

LAD interprets a decision table as a partially defined function $d = f(a_1, \ldots, a_n)$, in which case conditional attributes are called variables. The rows of the table describe the values of the variables for which the value of the function (*i.e.*, decision) is known. If all the attributes are binary, the decision table represents a partially defined Boolean function.

A typical problem in all three approaches is the problem of revealing functional dependencies between conditional attributes and the decision attribute. However, all three approaches use different terminology related to this problem.

In RS, a *super-reduct* is a set of conditional attributes that gives us the same information about the decision attribute as the whole set of conditional attributes. In other words, a super-reduct is a set of conditional attributes on which any two different rows (on the whole set of conditional attributes) with different decisions are different (based on conditional attributes from this set). A *reduct* is a minimal super-reduct, *i.e.*, a super-reduct not including any other super-reduct as its proper subset. In TT, the notion of *test* corresponds to the notion of super-reduct, and the notion of *dead-end test* corresponds to the notion of reduct. In LAD, the notion of *support set* corresponds to the notions of super-reduct and test.

In RS, a *decision rule* is a relation of the following form:

$$(a_{i_1} = b_1) \wedge \ldots \wedge (a_{i_m} = b_m) \rightarrow (d = b),$$

where $a_{i_1}, \ldots, a_{i_m}$ are conditional attributes, d is the decision attribute, and $b_1, \ldots, b_m, b$ are values of attributes $a_{i_1}, \ldots, a_{i_m}, d$, respectively. Here, we assume that the above decision rule is true in the table T, *i.e.*, each row of T having $b_1, \ldots, b_m$ at the intersection with the columns $a_{i_1}, \ldots, a_{i_m}$ is labeled with the decision b. In TT, decision rules are also called *representative tuples*. In LAD, the notion of *pattern* corresponds to the notion of decision rule. Decision rules are often generated on the basis of different kinds of reducts.

A *decision tree* for a given decision table T is a rooted directed tree in which nonterminal nodes are labeled with conditional attributes, terminal nodes are labeled with values of the decision attribute, and the edges coming out of any nonterminal node are labeled with pairwise different values of conditional attribute attached to this node. The computation of a decision tree on a given row is defined in a natural way. It is required that, for each row of the table T, the computation of the decision tree ends at a terminal node which is labeled with the decision attached to the considered row. The notion of decision tree is commonly used in TT and it also finds applications in RS and LAD. In TT, decision trees are also called *conditional tests*.

A common problem for TT, RS and LAD is the classification problem, *i.e.*, the problem of finding the value of the decision attribute based on the values of the conditional attributes. Tests (reducts, support sets), decision rules (patterns) and decision trees (conditional tests) are important tools for dealing with this problem. To solve the classification problem, we construct a *classification algorithm* (also known as *classifier, predictor, model*). In constructing classifiers, one can distinguish two main approaches.

In the first approach, which is typical for problems in computational geometry, discrete optimization, fault diagnosis, it is assumed that the decision table represents a complete description of the universe, in which case the efforts are focused on optimizing the classification algorithms in terms of time and space complexity.

In the second approach, which is typical for experimental and statistical data analysis, it is assumed that the decision table represents the universe only partially and the main task of the classifier is to predict the unseen part of the universe. In this case, the accuracy of prediction is, usually, more important than the complexity of the classifier, although the description length is an important issue in searching for high quality classifiers in all three approaches.

In addition to building a classifier, each of the three approaches has also developed numerous methods to solve a number of accompanying problems such as reduct generation, decision rule (pattern) generation, feature selection, discretization (binarization), symbolic value grouping, inducing classifiers, or clustering. In RS, the algorithms that solve these problems are often based on (approximate) Boolean reasoning, and in LAD, on optimization, combinatorics, and the theory of Boolean functions.

The book consists of the preface, three main parts devoted to TT, RS and LAD respectively, and final remarks. Each main part ends with a note about the founder of the corresponding theory.

Test Theory. The first part of the book, written by Igor Chikalov, Mikhail Moshkov and Beata Zielosko, is devoted to Test Theory (TT). This theory was created in the middle of fifties of the last century as a tool for solving problems of control and diagnosis of faults in circuits. In the middle of sixties, the methods of TT were extended to prediction problems in such domains as geology and medicine. This part consists of the chapter "Test Theory: Tools and Applications" and a note about the founder of TT – Sergei V. Yablonskii.

In the chapter, we consider three main areas of TT: (i) theoretical results related to tests, (ii) applications to control and diagnosis of faults, and (iii) applications to pattern recognition (prediction). We also discuss three less known areas of TT associated mainly with our research interests: the results of studies on infinite or large finite sets of attributes and applications to discrete optimization and mathematical linguistics. Also we give a common view on tests, decision trees and decision rule systems which can be placed at the intersection of TT and RS.

The chapter consists of seven sections.

The first three sections include theoretical results on tests, trees and rules. In the first section, we consider bounds on complexity and algorithms for construction of tests, rules and trees from, in some sense, uniform point of view. In the second section, we present results on the minimum length (cardinality) of tests and the number of reducts. This is the most well known area of the TT research. In the third section, we study problems over infinite or large finite sets of attributes. Such problems arise often in discrete optimization and computational geometry.

The following three sections deal with applications of TT. The fourth section is devoted to the TT methods for prediction. In the next two sections, we discuss applications of TT to problems with complete information, *i.e.*, when all possible tuples of attribute values for the considered problem are given. The fifth section is dedicated to the most developed area of TT applications, *i.e.*, control and diagnosis of faults. In the sixth section, we study problems of discrete optimization and recognition problems for words from formal languages. The last section contains conclusions.

Rough Sets. The second part of this book, written by Hung Son Nguyen and Andrzej Skowron, is dedicated to rough sets as a tool to deal with imperfect data, in particular, with vague concepts. In the development of rough set theory and its applications, one can distinguish three main stages. At the beginning, the researchers were concentrated on descriptive properties such as reducts of information systems preserving indiscernibility relations, description of concepts or classifications [8-10]. Next, they moved to applications of rough sets in machine learning, pattern recognition, and data mining. After gaining some experience, they developed the foundations for inductive reasoning leading to, *e.g.*, inducing classifiers. The first period was based on the assumption that objects are perceived by means of partial information represented by attributes. In the second period, it was also used the assumption that information about the approximated concepts is partial, too. Approximation spaces and searching strategies for relevant approximation spaces were recognized as the basic tools for rough sets. Important achievements both in theory and applications were obtained using Boolean reasoning and approximate Boolean reasoning applied, *e.g.*, in searching for relevant features, discretization, symbolic value grouping, or, in more general sense, in searching for relevant approximation spaces. Nowadays, we observe that a new period is emerging in which two new important topics are investigated: (i) strategies for discovering relevant (complex) contexts of analyzed objects or granules, which are strongly related to the information granulation process and granular computing, and (ii) interactive computations on granules. Both directions aim at developing tools for approximation of complex vague concepts such as behavioral patterns or adaptive strategies, making it possible to achieve the satisfactory qualities of the resulting interactive computations. This chapter presents this development from the rudiments of rough sets to some challenges.

In more details, the contents of this chapter are as follows. The chapter starts with a short discussion on vague concepts. Next, the basic concepts of rough set theory are recalled, including indiscernibility and discernibility relations, approximation of concepts, rough sets, decision rules, dependencies, reducts, discernibility and Boolean reasoning as the main methodology used in developing algorithms and heuristics based on rough sets, and also rough membership functions. Next, some extensions of the rough set approach are briefly presented. In the next part of this chapter, the relationship of the rough set approach to inductive reasoning is discussed. In particular, an outline of the rough set approach to inducing relevant approximation spaces and rough set based-classifiers, is given. Also, some comments on the relationship of the rough set approach and higher order vagueness is included. In the following part of the chapter, an extension of the rough set approach from concept approximation to approximation of ontologies is presented.

The rough set approach based on the combination of rough sets and Boolean reasoning to scalability in data mining is discussed in the following section of this chapter. Some comments on relationships of rough sets and logic are also included.

Finally, some challenging issues for rough sets are included in the last section of this chapter. Interactive Granular Computing (IGC), in particular, Interactive Rough Granular Computing (IRGC) are proposed as a framework that makes it possible to search for solutions to problems related to inducing of relevant contexts, process mining and perception based computing (PBC).

Logical Analysis of Data. The third part of the book was written by Vadim Lozin and Irina Lozina and is devoted to Logical Analysis of Data (LAD), the youngest of the three approaches. The idea of LAD was first described by Peter L. Hammer in a lecture given in 1986 at the International Conference on Multi-attribute Decision Making via OR-based Expert Systems [7] and was later expanded and developed in [6]. That first publication was followed by a stream of research studies. In early publications, the focus of research was on theoretical developments and on computational implementation. In recent years, attention was concentrated on practical applications varying from medicine to credit risk ratings. Following this pattern, we divided the chapter devoted to LAD into three main sections: Theory, Methodology and Applications.

In the section devoted to theory, we define the main notions used in Logical Analysis of Data, such as partially defined Boolean function, pattern and discuss various problems associated with these notions. This section is intended mainly for theoreticians and can be skipped, except possibly the first subsection devoted to terminology and notation, by those who are interested in practical implementations of the methodology.

The section devoted to methodology describes the main steps in Logical Analysis of Data, that include binarization, attribute selection, pattern generation, model construction and validation. We also outline specific algorithms implementing these steps.

In the section devoted to applications, we illustrate the LAD methodology with a number of particular examples, such as estimating passenger show rates in the airline industry and credit risk ratings.

We do hope that this book will stimulate intensive and successful research on the relationships between the three data analysis approaches as well as the development of new methods based on a combination of the existing methods.

Igor Chikalov, Vadim Lozin, Irina Lozina, Mikhail Moshkov
Hung Son Nguyen, Andrzej Skowron, Beata Zielosko

Coventry, Thuwal, Warsaw, January 2012

References

1. Blake, A.: Canonical Expressions in Boolean Algebra. Dissertation, Dept. of Mathematics, University of Chicago, 1937. University of Chicago Libraries (1938)
2. Boole, G.: The Mathematical Analysis of Logic. G. Bell, London (1847), (reprinted by Philosophical Library, New York, 1948)
3. Boole, G.: An Investigation of the Laws of Thought. Walton, London (1854), (reprinted by Dover Books, New York, 1954)
4. Brown, F.: Boolean Reasoning. Kluwer Academic Publishers, Dordrecht (1990)
5. Chegis, I. A., Yablonskii, S. V.: Logical methods of control of work of electric schemes: Trudy Mat. Inst. Steklov. **51**, 270–360 (1958), (in Russian)
6. Crama, Y., Hammer, Peter, L., Ibaraki, T.: Cause-effect relationships and partially defined Boolean functions. Annals of Operations Research **16**, 299–326 (1988)
7. Hammer, Peter, L.: Partially defined Boolean functions and cause-effect relationships. In: International Conference on Multi-attribute Decision Making via OR-based Expert Systems. University of Passau, Passau, Germany, April (1986)
8. Pawlak, Z.: Rough sets. Basic notions. ICS PAS Reports **431**/81. Institute of Computer Science Polish Academy of Sciences (ICS PAS), pp. 1–12, Warsaw, Poland (1981)
9. Pawlak, Z.: Rough sets. International Journal of Computer and Information Sciences **11**, 341–356 (1982)
10. Pawlak, Z.: Rough sets - Theoretical Aspects of Reasoning about Data. *System Theory, Knowledge Engineering and Problem Solving* **9**, Kluwer Academic Publishers, Boston, Dordrecht (1991)
11. Yablonskii, S. V., Chegis, I. A.: On tests for electric circuits. Uspekhi Mat. Nauk **10**, 182–184 (1955), (in Russian)

Acknowledgements

We are greatly indebted to King Abdullah University of Science and Technology (KAUST) and especially to Professor Jean M.J. Fréchet, Vice President of KAUST for Research and Professor David Keyes, Dean of Mathematical and Computer Sciences and Engineering Division of KAUST for various supports.

We extend an expression of gratitude to Professor Janusz Kacprzyk, to Dr. Thomas Ditzinger and to the Series Intelligent Systems Reference Library staff at Springer for their support in making this book possible.

Igor Chikalov, Vadim Lozin, Irina Lozina, Mikhail Moshkov
Hung Son Nguyen, Andrzej Skowron, Beata Zielosko

Coventry, Thuwal, Warsaw, January 2012

Contents

Part II: Rough Sets

Part III: Logical Analysis of Data

Part I

Test Theory

1

Test Theory: Tools and Applications

Test Theory (TT) was created by Yablonskii and Chegis in the middle of fifties of the last century as a tool for analysis of problems connected with control and diagnosis of faults in circuits. The first short paper [75] was published in 1955, and the principal paper [10] – in 1958. In 1966 Dmitriev, Zhuravlev and Krendelev [16] adapted test approach to the study of pattern recognition (prediction) problems. Later TT was developed in the two directions related to tools and applications.

Sergei V. Yablonskii was the main founder and inspirer of this theory until his death in 1998. This chapter is followed by a short note about his life and work.

The chapter consists of seven sections. The first three contain description of tools created in TT, the next three are devoted to TT applications, and the last one contains conclusions.

The main notions of TT are test table (decision table), test (super-reduct, support set), dead-end test (reduct), and conditional test (decision tree). Also the notions of representative set (decision rule, pattern) and nondeterministic decision tree (decision rules system) are used in TT.

We present tools created in TT for the study of decision tables, tests, reducts, decision trees, decision rules, and decision rule systems. We use the terminology which is common for publications in English with one exception: we use the term test instead of the term super-reduct.

The chapter begins (see Sect. 1.1) from a comparative study of tests, decision trees and decision rule systems. Relationships among these notions, bounds on complexity including common, and similar approaches for construction of tests, trees and rule systems are considered there.

The greatest efforts in TT were focused on studying tests. In Sect. 1.2, we consider bounds on the length (cardinality) of tests, and on the number of reducts, including results true for almost all decision tables.

Decision table is an adequate model for the case when finite and relatively small number of attributes can be used for problem solving. In Sect. 1.3, we study complexity of decision trees and decision rule systems over infinite or large finite sets of attributes.

I. Chikalov et al.: Three Approaches to Data Analysis, ISRL 41, pp. 3–61.
springerlink.com

All applications of TT can be divided into two groups: when complete information about the problem (*i.e.*, the whole decision table) is known and when we have incomplete information about the problem (*i.e.*, only some of rows of the whole decision table are known).

We start from the latter case. In Sect. 1.4, applications of TT to pattern recognition are described. In this case, only a part T of the initial decision table T' is known. The goal is, given the subtable T, design a classifier that predicts values of the decision attribute for rows from T'.

The next two sections deal with applications with complete information about corresponding decision tables.

Section 1.5 is devoted to the oldest area of applications, control and diagnosis of faults, with focus on contact networks and combinatorial circuits.

In Sect. 1.6, we consider two less known areas of applications of TT connected with discrete optimization and mathematical linguistics (problem of word recognition for formal languages).

1.1 Decision Trees, Rules and Tests

Decision tables and decision trees, rules and tests are popular ways of data and knowledge representation. Decision trees, decision rule systems and some structures based on tests can be considered as predictors or as algorithms for solving of problems given by finite sets of attributes.

In this section, we study tests, rules and trees from a general point of view which is common for TT and RS. We hope that such a presentation simplifies the understanding of the chapter.

This section contains three subsections. In Sect. 1.1.1, main notions are described as well as relationships among decision trees, rule systems and tests. Section 1.1.2 contains results related to bounds on complexity of decision trees, rules and tests. In Sect. 1.1.3, approximate and exact algorithms for construction and optimization of decision trees, rules and tests are presented.

The section is primarily based on monographs [13, 42, 44, 45].

1.1.1 Main Notions and Relationships

In this subsection, we present main notions connected with k-valued decision tables, decision trees, rules and tests.

Let k be a natural number such that $k \geq 2$. A (k-valued) *decision table* is a rectangular table which elements belong to the set $\{0,\ldots,k-1\}$. Columns of this table are labeled with conditional attributes $f_1,\ldots,f_n$. Rows of the table are pairwise different, and each row is labeled with a nonnegative integer – value of the decision attribute d. We denote by $N(T)$ the number of rows in the decision table T.

We associate *a game* of two players with this table. The first player chooses a row in the table and the second player should recognize a decision attached to this row. To this end, he can choose columns (attributes) and ask the first player what is at the intersection of the considered row and these columns.

A *decision tree over* T is a finite tree with root in which each terminal node is labeled with a decision (a nonnegative integer), and each nonterminal node (such nodes will be called *working*) is labeled with an attribute from the set $\{f_1,\ldots,f_n\}$. At most k edges start in each working node. These edges are labeled with pairwise different numbers from the set $\{0,\ldots,k-1\}$.

Let Γ be a decision tree over T. For a given row r of T, the computation of this tree starts from the root of Γ. If the considered node is terminal then the result of Γ computation is the number attached to this node. Let us assume that the currently considered node is a working node labeled with an attribute f_i. If the value of f_i in the considered row is equal to the number attached to one of edges outgoing from the node then we pass along this edge. Otherwise, the computation of Γ is finished in the considered nonterminal node without result.

We will say that Γ is a *decision tree for* T if for any row of T the computation of Γ ends at a terminal node, which is labeled with the decision attached to the considered row .

We denote by $h(\Gamma)$ the *depth* of Γ which is the maximum length of a path from the root to a terminal node. We denote by $h(T)$ the *minimum depth* of a decision tree for the table T.

We will consider also average depth of decision trees. To this end, we assume that a *probability distribution* $P=(p_1,\ldots,p_{N(T)})$ *for* T is given, where $p_1,\ldots,p_{N(T)}$ are positive real numbers and $\sum_{i=1}^{N(T)} p_i = 1$.

Then the *average depth of* Γ *relative to* P is equal to

$$h_{avg}(\Gamma,P) = \sum_{i=1}^{N(T)} p_i l_i \, ,$$

where l_i is the length of a path in Γ from the root to a terminal node in which the computation of Γ for the i-th row ends. We denote by $h_{avg}(T,P)$ the *minimum average depth* of a decision tree for the table T relative to P.

The value

$$H(P) = -\sum_{i=1}^{N(T)} p_i \log_2 p_i$$

is called the *entropy* of the probability distribution P.

A *decision rule over* T is an expression of the kind

$$f_{i_1} = b_1 \wedge \ldots \wedge f_{i_m} = b_m \to t \, ,$$

where $f_{i_1},\ldots,f_{i_m} \in \{f_1,\ldots,f_n\}$, $b_1,\ldots,b_m \in \{0,\ldots,k-1\}$, and t is a value of the decision attribute d. The number m is called the *length* of the rule. This rule is called *realizable* for a row $r=(\delta_1,\ldots,\delta_n)$ if

$$\delta_{i_1} = b_1, \ldots, \delta_{i_m} = b_m \, .$$

The rule is called *true* for T if for any row r of T, such that the rule is realizable for the row r, this row r is labeled with the value of decision attribute equal to t. We denote by $L(T,r)$ the *minimum length* of a rule over T which is true for T and realizable for r. We will say that the considered rule is a *rule for T and r* if this rule is true for T and realizable for r.

A *decision rule system S over T* is a nonempty finite set of rules over T. A system S is called a *complete decision rule system for T* if each rule from S is true for T, and for every row of T there exists a rule from S which is realizable for this row. We denote by $L(S)$ the *maximum length* of a rule from S, and by $L(T)$ we denote the *minimum value of $L(S)$* among all complete decision rule systems S for T. The parameter $L(S)$ will be called sometimes the *depth* of S.

A *test for T* is a subset of columns (conditional attributes) such that at the intersection with these columns any two rows with different decisions are different. The cardinality of a test will be called sometimes the *length* of this test. A *reduct for T* is a test for T for which each proper subset is not a test. It is clear that each test has a reduct as a subset. We denote by $R(T)$ the *minimum cardinality* of a reduct for T.

Now, we present some results connected with relationships among decision trees, rules and tests.

Theorem 1.1. *Let T be a decision table with n columns labeled with attributes $f_1,\dots,f_n$.*

1. *If Γ is a decision tree for T then the set of attributes attached to working nodes of Γ is a test for the table T.*
2. *Let $F = \{f_{i_1},\dots,f_{i_m}\}$ be a test for T. Then there exists a decision tree Γ for T which uses only attributes from F and for which $h(\Gamma) = m$.*

Corollary 1.2. *Let T be a decision table. Then $h(T) \le R(T)$.*

Theorem 1.3. *Let T be a decision table with n columns labeled with attributes $f_1,\dots,f_n$.*

1. *If S is a complete system of decision rules for T then the set of attributes from rules in S is a test for T.*
2. *If $F = \{f_{i_1},\dots,f_{i_m}\}$ is a test for T then there exists a complete system S of decision rules for T which uses only attributes from F and for which $L(S) = m$.*

Corollary 1.4. $L(T) \le R(T)$.

Let Γ be a decision tree for T and τ be a path in Γ from the root to a terminal node in which working nodes are labeled with attributes $f_{i_1},\dots,f_{i_m}$, edges are labeled with numbers $b_1,\dots,b_m$, and the terminal node of τ is labeled with the decision t. We associate with τ the decision rule

$$f_{i_1} = b_1 \wedge \dots \wedge f_{i_m} = b_m \to t\ .$$

Theorem 1.5. *Let Γ be a decision tree for T, and S be the set of decision rules corresponding to paths in Γ from the root to terminal nodes. Then S is a complete system of decision rules for T and $L(S) = h(\Gamma)$.*

Corollary 1.6. $L(T) \le h(T)$.

1.1.2 Lower and Upper Bounds

In this subsection, we present lower and upper bounds on the minimum depth and average depth of decision trees, cardinality (length) of tests and length of rules.

From Corollaries 1.2 and 1.6 it follows that $L(T) \leq h(T) \leq R(T)$. So, each lower bound on $L(T)$ is also a lower bound on $h(T)$ and $R(T)$, and each lower bound on $h(T)$ is also a lower bound on $R(T)$.

1.1.2.1 Lower Bounds

Now, we present some lower bounds on the value $h(T)$ and, consequently, on the value $R(T)$. We denote by $D(T)$ the number of different decisions in T.

Theorem 1.7. *Let T be a nonempty decision table. Then*

$$h(T) \geq \log_k D(T) .$$

Theorem 1.8. *Let T be a decision table. Then*

$$h(T) \geq \log_k((k-1)R(T)+1) .$$

A nonempty decision table T will be called a *diagnostic* table if rows of this table are labeled with pairwise different decisions. Note, that for a diagnostic table T the equality $D(T) = N(T)$ holds. Now, we present a lower bound on average depth of decision trees for diagnostic decision tables.

Theorem 1.9. *Let T be a diagnostic decision table and P be a probability distribution for T. Then*

$$h_{avg}(T,P) \geq \frac{H(P)}{\log_2 k} .$$

Example 1.10. Let us consider 2-valued decision table T depicted in Fig. 1.1. For this table $D(T) = 3$. Using Theorem 1.7 we obtain $h(T) \geq \log_2 3$, and therefore $h(T) \geq 2$. This table has exactly two tests: $\{f_1, f_2, f_3\}$ and $\{f_1, f_3\}$, so $R(T) = 2$. Using Theorem 1.8 we obtain $h(T) \geq \log_2 3$, and thus $h(T) \geq 2$.

In fact, $h(T) = 2$. A decision tree for the table T which depth is equal to 2 is depicted in Fig. 1.2.

Let T be a decision table with n columns labeled with attributes $f_1, \ldots, f_n$. A *subtable* of the table T is a table obtained from T by removal of some rows. Let $f_{i_1}, \ldots, f_{i_m} \in \{f_1, \ldots, f_n\}$ and $\delta_1, \ldots, \delta_m \in \{0, \ldots, k-1\}$. By $T(f_{i_1}, \delta_1) \ldots (f_{i_m}, \delta_m)$ we denote the subtable of the table T which consists of rows that at the intersection with columns $f_{i_1}, \ldots, f_{i_m}$ have numbers $\delta_1, \ldots, \delta_m$.

We will say that T is a *degenerate* table if T does not have rows or all rows of T are labeled with the same decision.

Now, we define a parameter $M(T)$ of the table T.

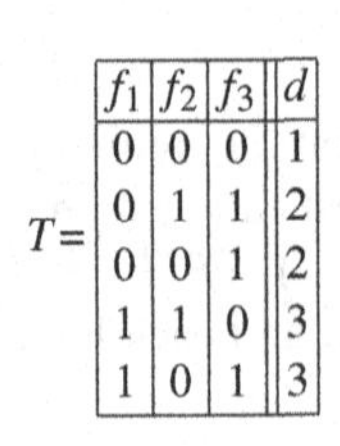

$T=$

f_1	f_2	f_3	d
0	0	0	1
0	1	1	2
0	0	1	2
1	1	0	3
1	0	1	3

Fig. 1.1.

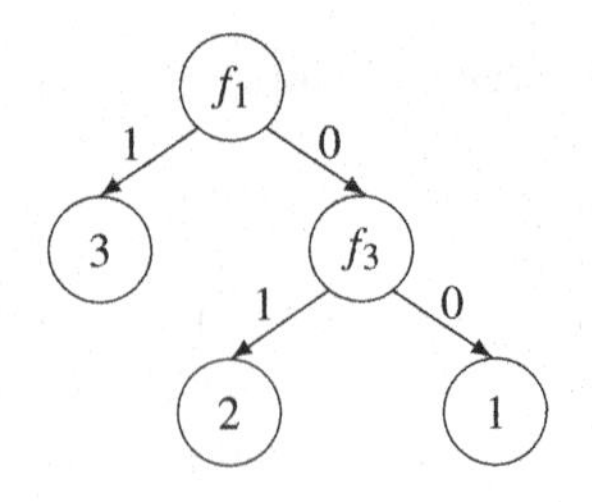

Fig. 1.2.

If T is degenerate then $M(T) = 0$. Let T be a nondegenerate. Let

$$\bar{\delta} = (\delta_1, \ldots, \delta_n) \in \{0, \ldots, k-1\}^n .$$

By $M(T, \bar{\delta})$ we denote the minimum natural m such that there exist $f_{i_1}, \ldots, f_{i_m} \in \{f_1, \ldots, f_n\}$ for which $T(f_{i_1}, \delta_{i_1}) \ldots (f_{i_m}, \delta_{i_m})$ is a degenerate table. Then

$$M(T) = \max\{M(T, \bar{\delta}) : \bar{\delta} \in \{0, \ldots, k-1\}^n\} .$$

We consider one more definition of the parameter $M(T, \bar{\delta})$. If $\bar{\delta}$ is a row of T then $M(T, \bar{\delta})$ is the minimum number of columns on which $\bar{\delta}$ is different from all rows with different decisions. Let $\bar{\delta}$ be not a row of T. Then $M(T, \bar{\delta})$ is the minimum number of columns on which $\bar{\delta}$ is different from all rows of T with the exception, possibly, of some rows with the same decision.

Note that the parameter $M(T)$ is close to the notion of *extended teaching dimension* [24, 25], and the parameter $L(T)$ (see Theorem 1.16) is close to the notion of *teaching dimension* [23].

Example 1.11. To find the value $M(T)$ for the decision table T depicted in Fig. 1.1, we need to find the value $M(T, \bar{\delta})$ for each $\bar{\delta} \in \{0, 1\}^3$. We obtain that $M(T) = 2$ (see Fig. 1.3).

$T=$

f_1	f_2	f_3	d	$M(T, \bar{\delta})$
0	0	0	1	2
0	1	1	2	2
0	0	1	2	2
1	1	0	3	1
1	0	1	3	1
1	1	1		1
1	0	0		1
0	1	0		2

Fig. 1.3.

In the definition of parameter $M(T)$, we use not only rows of T but also tuples $\bar{\delta}$ which are not rows of T. Example 1.12 helps to understand this definition.

Example 1.12. Let us consider 2-valued decision table T with n columns and n rows labeled with values of the decision attribute $1,\ldots,n$. For $i=1,\ldots,n$, the i-th row has 1 only at the intersection with the column f_i.

It is clear that for each row $\bar{\delta}$ of T the equality $M(T,\bar{\delta})=1$ holds. Let us consider the tuple $\bar{0}=(0,\ldots,0)\in\{0,1\}^n$. This tuple is not a row of T. It is easy to see that $M(T,\bar{0})=n-1$.

Theorem 1.13. *Let T be a decision table. Then*

$$h(T) \geq M(T) .$$

Theorems 1.7, 1.8, and 1.13 present three lower bounds on the depth of a decision tree for the table T. Unfortunately, sometimes each of these bounds is not enough precise. In this case we can use an approach based on the notion of proof-tree.

Example 1.14. Let us consider the problem of computation of the function

$$f(x_1,x_2,x_3)=x_1x_2 \vee x_1x_3 \vee x_2x_3$$

with the help of decision trees using values of variables x_1, x_2 and x_3. The corresponding decision table T is depicted in Fig. 1.4.

$T=$

x_1	x_2	x_3	f
0	0	0	0
0	0	1	0
0	1	0	0
0	1	1	1
1	0	0	0
1	0	1	1
1	1	0	1
1	1	1	1

Fig. 1.4.

Our aim is to evaluate $h(T)$. Note that the value of the considered function (the function of voting) on a tuple $(\delta_1,\delta_2,\delta_3)$ is equal to 0 if the number of 0 among δ_1, δ_2, and δ_3 is maximum, and it is equal to 1 if the number of 1 among δ_1, δ_2, and δ_3 is maximum.

It is clear that $D(T)=2$. From Theorem 1.7 it follows that $h(T)\geq 1$. One can show that $R(T)=3$. From Theorem 1.8 it follows that $h(T)\geq 2$. It is not difficult to see that for any $\bar{\delta}\in\{0,1\}^3$ the equality $M(T,\bar{\delta})=2$ holds. Hence, $M(T)=2$. From Theorem 1.13 it follows that $h(T)\geq 2$. Thus, we have the following lower bound: $h(T)\geq 2$. But it is impossible, for decision table depicted in Fig. 1.4, construct a decision tree for T which depth is equal to 2.

In such situation we should find a way to obtain more precise lower bound. In the considered case we can use the following reasoning: if the first question is, for example, about the value of x_1, then we will answer that $x_1 = 0$; if the second question is, for example, about the value of x_2, then we will answer that $x_2 = 1$. Thus, it will be necessary to ask the third question about the value of x_3.

In some sense we are saying about a *strategy of the first player* in the game which is *modified* in the following way: the first player does not choose a row at the beginning of the game, but at least one row must satisfy his answers on questions asked by the second player.

A strategy of the first player is depicted in Fig. 1.5.

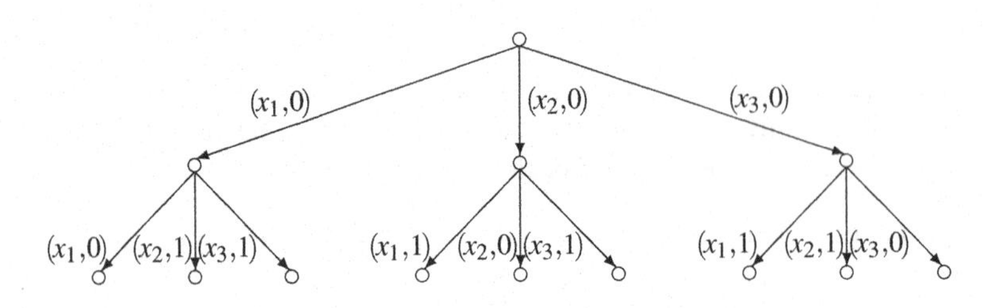

Fig. 1.5.

We can see that if the first player uses this strategy, the second player after any two questions will localize the considered row in a table which is not degenerate. Therefore he should make additional third step, and hence, $h(T) \geq 3$.

Now we consider the notion of strategy of the first player more formally. We will say about so-called proof-trees.

Let T be a decision table with n columns labeled with attributes $f_1, \dots, f_n$, m be a natural number, and $m \leq n$.

A (T,m)-*proof-tree* is a finite directed tree G with the root in which the length of each path from the root to a terminal node is equal to $m-1$. Nodes of this tree are not labeled. In each nonterminal node exactly n edges start. These edges are labeled with pairs of the kind $(f_1, \delta_1), \dots, (f_n, \delta_n)$ respectively where $\delta_1, \dots, \delta_n \in \{0, \dots, k-1\}$. For example, in Fig. 1.5 a $(T,3)$-proof-tree is depicted, where T is the table depicted in Fig. 1.4.

Let v be an arbitrary terminal node of G and $(f_{i_1}, \delta_1), \dots, (f_{i_{m-1}}, \delta_{m-1})$ be pairs attached to edges in the path from the root of G to the terminal node v. Denote $T(v) = T(f_{i_1}, \delta_1) \dots (f_{i_{m-1}}, \delta_{m-1})$.

We will say that G is a *proof-tree for the bound* $h(T) \geq m$ if for any terminal node v the subtable $T(v)$ is not degenerate.

Theorem 1.15. *Let T be a nondegenerate decision table with n columns and m be a natural number such that $m \leq n$. Then a proof-tree for the bound $h(T) \geq m$ exists if and only if the inequality $h(T) \geq m$ holds.*

So, the approach based on proof-trees can give the best lower bound on the value $h(T)$ for every nondegenerate decision table T.

Now, we present exact formula for $L(T)$ in terms of parameters $M(T, \bar{\delta})$.

Theorem 1.16. *Let T be a decision table and $\Delta(T)$ be the set of rows of T. Then $L(T,\bar{\delta}) = M(T,\bar{\delta})$ for any row $\bar{\delta} \in \Delta(T)$ and $L(T) = \max\{M(T,\bar{\delta}) : \bar{\delta} \in \Delta(T)\}$.*

Corollary 1.17. $L(T) \leq M(T) \leq h(T) \leq R(T)$.

Example 1.18. For the table T_1 depicted in Fig. 1.1 we have $\max\{M(T_1,\bar{\delta}) : \bar{\delta} \in \Delta(T_1)\} = 2$ (see Example 1.11). Therefore $L(T_1) = 2$.

For the decision table T_2 depicted in Fig. 1.4 we have $\max\{M(T_2,\bar{\delta}) : \bar{\delta} \in \Delta(T_2)\} = 2$ (see Example 1.14). Therefore $L(T_2) = 2$.

1.1.2.2 Upper Bounds

First, we present an upper bound on the value $R(T)$. It will be also an upper bound on the values $L(T)$ and $h(T)$.

Theorem 1.19. *Let T be a decision table. Then*

$$R(T) \leq N(T) - 1 .$$

Example 1.20. Let n be a natural number. We consider a 2-valued decision table T_n which contains n columns labeled with conditional attributes $f_1,\dots,f_n$ and $n+1$ rows. For $i = 1,\dots,n$, the i-th row has 1 only at the intersection with the column f_i. This row is labeled with the decision 1. The last $(n+1)$-th row is filled by 0 only and is labeled with the decision 2. One can show that $N(T_n) = n+1$ and $R(T_n) = n$. Thus, the bound from Theorem 1.19 is unimprovable in the general case.

Example 1.21. Let T be the table depicted in Fig. 1.1. We know (see Example 1.10) that $R(T) = 2$. Theorem 1.19 gives us the upper bound $R(T) \leq 4$.

Now we present an upper bound on the value $h(T)$. It will be also an upper bound on $L(T)$.

Theorem 1.22. *Let T be a decision table. Then*

$$h(T) \leq M(T) \log_2 N(T) .$$

It is possible to improve the bound from Theorem 1.22 and show that

$$h(T) \leq \begin{cases} M(T), & \text{if } M(T) \leq 1 , \\ 2\log_2 N(T) + M(T), & \text{if } 2 \leq M(T) \leq 3 , \\ \frac{M(T)\log_2 N(T)}{\log_2 M(T)} + M(T), & \text{if } M(T) \geq 4 . \end{cases}$$

Corollary 1.23. *Let T be a diagnostic decision table. Then*

$$\max\{M(T), \log_k N(T)\} \leq h(T) \leq M(T) \log_2 N(T) .$$

In the next statement, we present an upper bound on the minimum average depth of a decision tree for given decision table and probability distribution.

Theorem 1.24. *Let T be a nonempty decision table and P be a probability distribution for T. Then*

$$h_{avg}(T,P) \le \begin{cases} M(T), & \text{if } M(T) \le 1\,, \\ M(T)+2H(P), & \text{if } 2 \le M(T) \le 3\,, \\ M(T)+\frac{M(T)}{\log_2 M(T)}H(P), & \text{if } M(T) \ge 4\,. \end{cases}$$

1.1.3 Algorithms

In this subsection, we present algorithms for construction and optimization of tests, decision rules and trees. We consider the following problems of optimization:

- *Problem of minimization of test cardinality*: for a given decision table T, it is required to construct a test for this table, which has minimum cardinality.
- *Problem of minimization of decision rule length*: for a given decision table T and row r of T, it is required to construct a decision rule over T which is true for T, realizable for r, and has minimum length.
- *Problem of optimization of decision rule system*: for a given decision table T, it is required to construct a complete decision rule system S for T with minimum value of parameter $L(S)$.
- *Problem of minimization of decision tree depth*: for a given decision table T, it is required to construct a decision tree for this table which has minimum depth.

We show that these problems are NP-hard and discuss, for these problems, some results on precision of approximate polynomial algorithms. We present greedy algorithms for these problems and exact algorithms, based on dynamic programming approach, for optimization of decision trees and rules.

First, we consider well known set cover problem.

1.1.3.1 Set Cover Problem

Let A be a set containing $N > 0$ elements, and $F = \{S_1,\dots,S_p\}$ be a family of subsets of the set A such that $A = \bigcup_{i=1}^{p} S_i$. A subfamily $\{S_{i_1},\dots,S_{i_t}\}$ of the family F is called a *cover* if $\bigcup_{j=1}^{t} S_{i_j} = A$. The problem of searching for a cover with minimum cardinality t is called the *set cover problem*. It is well known that this problem is an NP-hard problem. U. Feige [21] proved that if $NP \not\subseteq DTIME(n^{O(\log\log n)})$ then for any ε, $0 < \varepsilon < 1$, there is no polynomial algorithm that constructs a cover which cardinality is at most $(1-\varepsilon)C_{\min}\ln N$ where $C_{\min}$ is the minimum cardinality of a cover.

We now consider well known *greedy* algorithm for set cover problem.

Set $B := A$, and $COVER := \emptyset$.

(*) In the family F we find a set S_i with minimum index i such that

$$|S_i \cap B| = \max\{|S_j \cap B| : S_j \in F\}\,.$$

Then we set $B := B \setminus S_i$ and $COVER := COVER \cup \{S_i\}$. If $B = \emptyset$ then we finish the computation of the algorithm. The set $COVER$ is the result of the algorithm computation. If $B \neq \emptyset$ then we return to the label (*).

We denote by C_{greedy} the cardinality of the cover constructed by greedy algorithm, and by $C_{\min}$ – the minimum cardinality of a cover.

Theorem 1.25. $C_{\text{greedy}} \leq C_{\min} \ln N + 1$.

The considered (or very similar) bound was obtained independently by different authors: by Nigmatullin [48], Johnson [26], etc. It was improved by Slavík in [62, 63]:

$$C_{\text{greedy}} \leq C_{\min}(\ln N - \ln \ln N + 1) .$$

Also Slavík has shown that it is impossible to improve this bound essentially.

Using the mentioned result of Feige we obtain that if $NP \not\subseteq DTIME\,(n^{O(\log \log n)})$ then the greedy algorithm is close to the best (from the point of view of precision) approximate polynomial algorithms for solving the set cover problem.

1.1.3.2 Greedy Algorithm for Test Construction

We can use the greedy algorithm for set cover problem to construct a test for given decision table T.

Let T be a decision table containing n columns labeled with attributes $f_1, \ldots, f_n$. We consider a set cover problem $A(T), F(T) = \{S_1, \ldots, S_n\}$ where $A(T)$ is the set of all unordered pairs of rows of the table T with different decisions. For $i = 1, \ldots, n$, the set S_i coincides with the set of all pairs of rows from $A(T)$ which are different in the column f_i. One can show that the set of columns $\{f_{i_1}, \ldots, f_{i_m}\}$ is a test for the table T iff the subfamily $\{S_{i_1}, \ldots, S_{i_m}\}$ is a cover for the set cover problem $A(T)$, $F(T)$.

We denote by $P(T)$ the number of unordered pairs of rows of T which have different decisions. It is clear that $|A(T)| = P(T)$. It is clear also that for the considered set cover problem $C_{\min} = R(T)$.

When we apply the greedy algorithm to the considered set cover problem, this algorithm constructs a cover which corresponds to a test for the table T. From Theorem 1.25 it follows that the cardinality of this test is at most

$$R(T) \ln P(T) + 1 .$$

We denote by $R_{\text{greedy}}(T)$ the cardinality of the test constructed by the following algorithm: for a given decision table T we construct the set cover problem $A(T), F(T)$ and then apply to this problem the greedy algorithm for set cover problem. According to what has been said we have the following statement.

Theorem 1.26. *Let T be a nondegenerate decision table. Then*

$$R_{\text{greedy}}(T) \leq R(T) \ln P(T) + 1 .$$

Example 1.27. Let us apply the considered algorithm to the table T depicted in Fig. 1.1. For this table, $A(T) = \{(1,2),(1,3),(1,4),(1,5),(2,4),(2,5),(3,4),(3,5)\}$ (we are writing here pairs of numbers of rows instead of pairs of rows), $F(T) = \{S_1,S_2,S_3\}$, $S_1 = \{(1,4),(1,5),(2,4),(2,5),(3,4),(3,5)\}$, $S_2 = \{(1,2),(1,4),(2,5),(3,4)\}$, and $S_3 = \{(1,2),(1,3),(1,5),(2,4),(3,4)\}$. At the first step, the greedy algorithm chooses S_1, and at the second step this algorithm chooses S_3. The constructed cover is $\{S_1,S_3\}$. The corresponding test is equal to $\{f_1,f_3\}$. As we know, this is the reduct with minimum cardinality.

It is not difficult to describe a polynomial time reduction of an arbitrary set cover problem to the problem of minimization of test cardinality for a decision table and prove the following two statements.

Proposition 1.28. *The problem of minimization of test cardinality is NP-hard.*

Theorem 1.29. *If $NP \not\subseteq DTIME(n^{O(\log\log n)})$ then for any ε, $0 < \varepsilon < 1$, there is no polynomial algorithm that for a given nondegenerate decision table T constructs a test for T which cardinality is at most*

$$(1-\varepsilon) R(T) \ln P(T) .$$

1.1.3.3 Greedy Algorithm for Decision Rule Construction

We can apply the greedy algorithm for set cover problem to construct decision rules.

Let T be a nondegenerate decision table containing n columns labeled with attributes $f_1,\ldots,f_n$, and $r = (b_1,\ldots,b_n)$ be a row of T labeled with a value of decision attribute equal to t. We consider a set cover problem $A(T,r)$, $F(T,r) = \{S_1,\ldots,S_n\}$ where $A(T,r)$ is the set of all rows of T with decisions different from t. For $i = 1,\ldots,n$, the set S_i coincides with the set of all rows from $A(T,r)$ which are different from r in the column f_i. One can show that the decision rule

$$f_{i_1} = b_{i_1} \wedge \ldots \wedge f_{i_m} = b_{i_m} \to t$$

is true for T (it is clear that this rule is realizable for r) if and only if the subfamily $\{S_{i_1},\ldots,S_{i_m}\}$ is a cover for the set cover problem $A(T,r)$, $F(T,r)$.

We denote by $P(T,r)$ the number of rows of T with decisions different from t. It is clear that $|A(T,r)| = P(T,r)$. It is clear also that for the considered set cover problem $C_{\min} = L(T,r)$.

When we apply the greedy algorithm to the considered set cover problem this algorithm constructs a cover which corresponds to a decision rule that is true for T and realizable for r. From Theorem 1.25 it follows that the length of this decision rule is at most

$$L(T,r) \ln P(T,r) + 1 .$$

We denote by $L_{\text{greedy}}(T,r)$ the length of the rule constructed by the following polynomial algorithm: for a given decision table T and row r of T we construct the set cover problem $A(T,r)$, $F(T,r)$ and then apply to this problem greedy algorithm for set cover problem. According to what has been said above we have the following statement.

Theorem 1.30. *Let T be a nondegenerate decision table and r be a row of T. Then*

$$L_{\text{greedy}}(T,r) \leq L(T,r) \ln P(T,r) + 1 .$$

Example 1.31. Let us apply the considered algorithm to the table T depicted in Fig. 1.1 and to the first row of this table. For $i = 1,\ldots,5$, we denote by r_i the i-th row of T. We have $A(T,r_1) = \{r_2,r_3,r_4,r_5\}$, $F(T,r_1) = \{S_1,S_2,S_3\}$, $S_1 = \{r_4,r_5\}$, $S_2 = \{r_2,r_4\}$, $S_3 = \{r_2,r_3,r_5\}$. At the first step, the greedy algorithm chooses S_3, and at the second step this algorithm chooses S_1. The constructed cover is $\{S_1,S_3\}$. The corresponding decision rule $f_1 = 0 \wedge f_3 = 0 \to 1$ has minimum length among rules over T that are true for T and realizable for r_1.

We can use the considered algorithm to construct a complete decision rule system for T. To this end we apply this algorithm sequentially to the table T and to each row r of T. As a result we obtain a system of rules S in which each rule is true for T and for every row of T there exists a rule from S which is realizable for this row.

We denote $L_{\text{greedy}}(T) = L(S)$ and $K(T) = \max\{P(T,r) : r \in \Delta(T)\}$, where $\Delta(T)$ is the set of rows of T. It is clear that $L(T) = \max\{L(T,r) : r \in \Delta(T)\}$. Using Theorem 1.30 we obtain

Theorem 1.32. *Let T be a nondegenerate decision table. Then*

$$L_{\text{greedy}}(T) \leq L(T) \ln K(T) + 1 .$$

Example 1.33. Let us apply the considered algorithm to the table T depicted in Fig. 1.1. As a result we obtain the following complete decision rules system for T:

$$S = \{f_1 = 0 \wedge f_3 = 0 \to 1, f_1 = 0 \wedge f_2 = 1 \to 2,$$
$$f_1 = 0 \wedge f_3 = 1 \to 2, f_1 = 1 \to 3, f_1 = 1 \to 3\} .$$

For this system, $L(S) = 2$. We know that $L(T) = 2$ (see Example 1.18).

One can prove the following four statements.

Proposition 1.34. *The problem of minimization of decision rule length is NP-hard.*

Theorem 1.35. *If $NP \not\subseteq DTIME(n^{O(\log\log n)})$ then for any ε, $0 < \varepsilon < 1$, there is no polynomial algorithm that for a given nondegenerate decision table T and row r of T constructs a decision rule which is true for T, realizable for r, and which length is at most*

$$(1-\varepsilon) L(T,r) \ln P(T,r) .$$

Proposition 1.36. *The problem of optimization of decision rule system is NP-hard.*

Theorem 1.37. *If $NP \not\subseteq DTIME(n^{O(\log\log n)})$ then for any ε, $0 < \varepsilon < 1$, there is no polynomial algorithm that for a given nondegenerate decision table T constructs a complete decision rule system S for T such that*

$$L(S) \leq (1-\varepsilon)L(T)\ln K(T) .$$

1.1.3.4 Greedy Algorithm for Decision Tree Construction

Now we describe an algorithm U which for a decision table T constructs a decision tree $U(T)$ for the table T. Let T have n columns labeled with attributes $f_1,\ldots,f_n$.

Step 1: Construct a tree consisting of a single node labeled with the table T and proceed to the second step.

Suppose $s \geq 1$ steps have been made already. The tree obtained at the step t will be denoted by G.

Step $(s+1)$: If no one node of the tree G is labeled with a table then we denote by $U(T)$ the tree G. The computation of the algorithm U is completed.

Otherwise, we choose certain node v in the tree G which is labeled with a subtable of the table T. Let the node v be labeled with the table T'. If T' is a degenerate table (all rows of the table are labeled with the same value t of the decision attribute) then instead of T' we mark the node v by the number t and proceed to the step $(s+2)$. Let T' be a nondegenerate table. Then, for $i = 1,\ldots,n$, we compute the value

$$Q(f_i) = \max\{P(T'(f_i,0)),\ldots,P(T'(f_i,k-1))\} .$$

We mark the node v by the attribute f_{i_0} where i_0 is the minimum i for which $Q(f_i)$ has minimum value. For each $\delta \in \{0,\ldots,k-1\}$ such that the subtable $T'(f_{i_0},\delta)$ is nonempty, we add to the tree G the node $v(\delta)$, mark this node by the table $T'(f_{i_0},\delta)$, draw the edge from v to $v(\delta)$, and mark this edge by δ. Proceed to the step $(s+2)$.

Example 1.38. Let us apply the algorithm U to the decision table T depicted in Fig. 1.1. After the first step, we obtain the tree which has only one node v that is labeled with the table T. The table T is not degenerate. So, for $i = 1,2,3$, we compute the value

$$Q(f_i) = \max\{P(T(f_i,0)),P(T(f_i,1))\} .$$

It is not difficult to see that $Q(f_1) = \max\{2,0\} = 2$, $Q(f_2) = \max\{3,1\} = 3$, and $Q(f_3) = \max\{1,2\} = 2$. It is clear that 1 is the minimum index for which the value of $Q(f_1)$ is minimum. So, after the second step we will have the tree G depicted in Fig. 1.6. We omit next steps. One can show that as a result of the algorithm U computation for the table T, we obtain the tree $U(T)$ depicted in Fig. 1.2.

Now, we evaluate the number of steps which the algorithm U makes during the construction of the decision tree $U(T)$.

Theorem 1.39. *Let T be a decision table. Then during the construction of the tree $U(T)$ the algorithm U makes at most $2N(T)+1$ steps.*

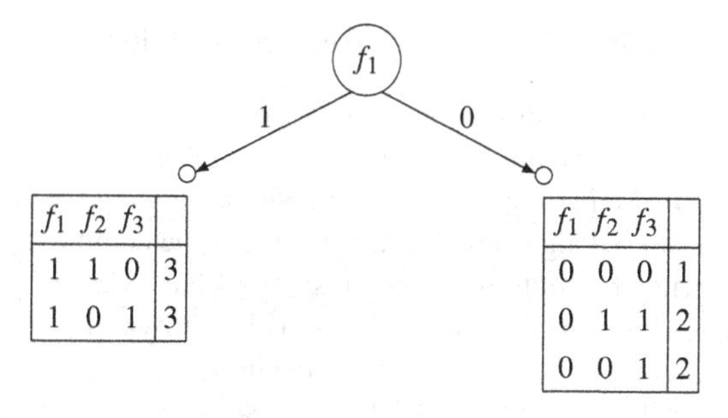

f_1	f_2	f_3	
1	1	0	3
1	0	1	3

f_1	f_2	f_3	
0	0	0	1
0	1	1	2
0	0	1	2

Fig. 1.6.

From here it follows that the algorithm U has polynomial time complexity. Now, we evaluate the accuracy of the algorithm U (as an algorithm for minimization of decision tree depth).

Theorem 1.40. *Let T be a nondegenerate decision table. Then*

$$h(U(T)) \leq M(T) \ln P(T) + 1 .$$

Using Theorem 1.13 we obtain the following

Corollary 1.41. *For any nondegenerate decision table T*

$$h(U(T)) \leq h(T) \ln P(T) + 1 .$$

It is possible to improve the considered bounds and show that for any decision table T

$$h(U(T)) \leq \begin{cases} M(T), & \text{if } M(T) \leq 1 , \\ M(T)(\ln P(T) - \ln M(T) + 1), & \text{if } M(T) \geq 2 , \end{cases}$$

and

$$h(U(T)) \leq \begin{cases} h(T), & \text{if } h(T) \leq 1 , \\ h(T)(\ln P(T) - \ln h(T) + 1), & \text{if } h(T) \geq 2 . \end{cases}$$

The last two bounds do not allow for essential improvement (see details in [38]).

It is not difficult to describe a polynomial time reduction of an arbitrary set cover problem to the problem of minimization of decision tree depth for a decision table, and prove the following two statements.

Proposition 1.42. *The problem of minimization of decision tree depth is NP-hard.*

Theorem 1.43. *If $NP \not\subseteq DTIME(n^{O(\log \log n)})$ then for any $\varepsilon > 0$ there is no polynomial algorithm which for a given nondegenerate decision table T constructs a decision tree for T which depth is at most $(1-\varepsilon)h(T)\ln P(T)$.*

Based on Theorems 1.26 and 1.29, 1.30 and 1.35, 1.32 and 1.37, 1.40 (Corollary 1.41) and 1.43 we can see that, under the assumption $NP \not\subseteq DTIME(n^{O(\log \log n)})$, the presented greedy algorithms are close (from the point of view of precision) to the best polynomial approximate algorithms for the minimization of test cardinality, rule length, tree depth and also, for the optimization of decision rule systems.

1.1.3.5 Exact Algorithm for Decision Tree Optimization

Now, we present exact algorithm for optimization of decision trees. This algorithm is based on dynamic programming approach and, for a given decision table, constructs a decision tree for this table with minimum depth (see also [43]). Of course, in the worst case the considered algorithm has exponential time complexity. However, we know that for restricted infinite information systems (see Sect. 1.3) this algorithm (for decision tables over the considered system) has polynomial time complexity depending on the number of columns (attributes) in the table.

The idea of algorithm is simple. Let T be a decision table with n columns labeled with attributes $f_1,\ldots,f_n$. If T is a degenerate decision table which rows are labeled with the same decision t, the tree which has exactly one node labeled with t, is an optimal decision tree for T. Let T be a nondegenerate table. In this case, $h(T)\geq 1$, and the root of any optimal decision tree for the table T is labeled with an attribute. We denote by $E(T)$ the set of attributes of T which are not constant, i.e., at the intersection with the column f_i they have at least two different values. We should consider only attributes from $E(T)$. For $f_i\in E(T)$, we denote by $E(T,f_i)$ the set of values of f_i in T.

We denote by $h(T,f_i)$ the minimum depth of a decision tree for the table T in which the root is labeled with the attribute $f_i\in E(T)$. It is clear that

$$h(T,f_i)=1+\max\{h(T(f_i,\delta)):\delta\in E(T,f_i)\}\,.$$

So, if we know values of $h(T(f_i,\delta))$ for any $f_i\in E(T)$ and $\delta\in E(T,f_i)$, we can find the value

$$h(T)=\min\{h(T,f_i):f_i\in E(T)\}$$

and at least one attribute f_i for which $h(T)=h(T,f_i)$.

Let for any $\delta\in E(T,f_i)$, Γ_δ be a decision tree for $T(f_i,\delta)$ with minimum depth. Then the following decision tree Γ is an optimal decision tree for T: the root of Γ is labeled with the attribute f_i and, for every $\delta\in E(T,f_i)$, an edge marked by δ issues from the root of Γ and enters the root of the tree Γ_δ. So, if we know optimal decision trees for subtables of the table T we can construct an optimal decision tree for the table T.

We now describe the algorithm for minimization of decision tree depth more formally. We denote this algorithm by W. A nonempty subtable T' of the table T will be called a *separable* subtable of T if there exist attributes $f_{i_1},\ldots,f_{i_m}$ from $\{f_1,\ldots,f_n\}$ and numbers $\delta_1,\ldots,\delta_m$ from $\{0,\ldots,k-1\}$ such that $T'=T(f_{i_1},\delta_1)\ldots(f_{i_m},\delta_m)$. We denote by $SEP(T)$ the set of all separable subtables of the table T including T.

The first part of the algorithm W computation is the construction of the set $SEP(T)$.

Step 1: We set $SEP(T)=\{T\}$ and pass to the second step. After the first step T is not labeled as a treated table.

Suppose $s\geq 1$ steps have been made already.

Step $(s+1)$: Let all tables in the set $SEP(T)$ be labeled as treated tables. In this case, we finish the first part of the algorithm W computation. Let there exist a table

$D \in SEP(T)$ which is not treated. We add to the set $SEP(T)$ all subtables of the kind $D(f_i, \delta)$, where $f_i \in E(D)$ and $\delta \in E(D, f_i)$, which were not in $SEP(T)$, mark the table D as treated and pass to the step $(s+2)$.

It is clear that during the first part the algorithm W makes exactly $|SEP(T)|+2$ steps.

The second part of the algorithm W computation is the construction of an optimal decision tree $W(T)$ for the table T. Beginning with smallest subtables from $SEP(T)$, the algorithm W at each step will correspond to a subtable from $SEP(T)$ an optimal decision tree for this subtable.

Suppose that $p \geq 0$ steps of the second part of algorithm W have been made already.

Step $(p+1)$: If the table T in the set $SEP(T)$ is labeled with a decision tree then this tree is the result of the algorithm W computation (we denote this tree by $W(T)$). Otherwise, choose in the set $SEP(T)$ a table D satisfying the following conditions:

a) the table D is not labeled with a decision tree;
b) either D is a degenerate table, or a nondegenerate table such that all separable subtables of D of the kind $D(f_i, \delta)$, $f_i \in E(D)$, $\delta \in E(D, f_i)$, are labeled with decision trees.

Let D be a degenerate table in which all rows are labeled with the same value of the decision attribute equal to t. Then we mark the table D by the decision tree consisting of one node which is labeled with the number t.

Otherwise, for any $f_i \in E(D)$, we construct a decision tree $\Gamma(f_i)$. The root of this tree is labeled with the attribute f_i. For each $\delta \in E(D, f_i)$, the root is the initial node of an edge which is labeled with the number δ. This edge enters the root of decision tree attached to the table $D(f_i, \delta)$. Mark the table D by one of the trees $\Gamma(f_j)$, $f_j \in E(D)$, having minimum depth, and proceed to the step $(p+2)$.

It is clear that during the second part, the algorithm W makes exactly $|SEP(T)+1|$ steps.

Theorem 1.44. *For any nondegenerate decision table T the algorithm W constructs a decision tree $W(T)$ for the table T such that $h(W(T)) = h(T)$, and makes exactly $2|SEP(T)|+3$ steps. The time of the algorithm W computation is bounded from below by $|SEP(T)|$, and bounded from above by a polynomial on $|SEP(T)|$ and on the number of columns in the table T.*

Similar algorithms exist also for other complexity measures, for example, for average depth of decision trees and number of nodes in decision trees [1, 11, 12, 13, 15]. Dynamic programming approach for optimization of decision trees was considered also in [2, 14].

1.1.3.6 Exact Algorithm for Decision Rule Optimization

We now describe an algorithm V for minimization of the length of decision rules (see also [78]). The algorithm V computation consists of two parts. The first part is connected with the construction of the set $SEP(T)$. This part coincides with the first part of the algorithm W. Now, we describe the second part of the algorithm V computation.

The second part is connected with the construction of an optimal decision rule $V(T,r)$ for each row r of T. Beginning with the smallest subtables from $SEP(T)$, the algorithm V at each step corresponds to each row r of a subtable T' an optimal decision rule for T' and r (an *optimal* rule for T' and r means a decision rule with minimum length which is true for T' and realizable for r).

Suppose $p \geq 0$ steps of the second part of algorithm V have been made already.

Step $(p+1)$. If each row r of the table T is labeled with a decision rule, then the rule attached to r is the result of the computation of V for T and r (we denote this rule by $V(T,r)$). Otherwise, choose in the set $SEP(T)$ a table D satisfying the following conditions:

a) rows of D are not labeled with decision rules;
b) either D is a degenerate table, or a nondegenerate table such that for all separable subtables of D of the kind $D(f_i,\delta)$, $f_i \in E(D)$, $\delta \in E(D,f_i)$, each row is labeled with a decision rule.

Let D be a degenerate table in which all rows are labeled with the same value of decision attribute equal to t. Then we mark each row of D by the decision rule $\to t$.

Let D be a nondegenerate decision table and $r = (\delta_1,\ldots,\delta_n)$ be a row of D labeled with a decision t. For any $f_i \in E(D)$ we construct a rule rule(r,f_i). Let the row r in the table $D(f_i,\delta_i)$ be labeled with the rule $\alpha_i \to t$. Then the rule rule(r,f_i) is equal to $f_i = \delta_i \wedge \alpha_i \to t$. We mark the row r of the table D with one of the rules rule(r,f_i), $f_i \in E(D)$, with minimum length. We denote this rule by $V(T,r)$. We attach a rule to each row of D in the same way, and proceed to the step $(p+2)$. It is clear that during the second part the algorithm V makes exactly $|SEP(T)|+1$ steps.

Theorem 1.45. *For any nondegenerate decision table T and any row r of T the algorithm V constructs a decision rule $V(T,r)$ which is true for T, realizable for r, and has minimum length $L(T,r)$. During the construction of optimal rules for rows of T, the algorithm V makes exactly $2|SEP(T)|+3$ steps. The time of the algorithm V computation is bounded from below by $|SEP(T)|$, and bounded from above by a polynomial on $|SEP(T)|$ and on the number of columns in the table T.*

Similar algorithms exist also for other cost functions, for example, coverage (see [3, 4]).

1.1.3.7 Optimization of Tests

It would be very interesting to have similar algorithm for the problem of minimization of test cardinality. However, if $P \neq NP$, then such algorithm does not exist.

Theorem 1.46. *If $P \neq NP$, then there is no algorithm which for a given decision table T constructs a test for T with minimum cardinality, and for which the time of computation is bounded from above by a polynomial depending on the number of columns in T and the number of separable subtables of T.*

1.2 Tests

Tests and dead-end tests (reducts) are the most studied objects in test theory. In this section, we consider results for tests and reducts obtained in TT.

The section consists of four subsections. In Sect. 1.2.1, main definitions are repeated. Section 1.2.2 contains results related to the length of tests and Sect. 1.2.3 – to the number of reducts. In Sect. 1.2.4, some algorithms for test construction are discussed.

1.2.1 Definitions

A *decision table* T is a rectangular table which elements belong to the set $\{0,...,k-1\}$ for some $k \geq 2$. Often we will consider *binary* decision tables, for which $k = 2$. Columns of this table are labeled with attributes $f_1,\ldots,f_n$. Rows of the table are pairwise different, and each row is labeled with a nonnegative integer (a decision). A decision table is called *nondegenerate* if there is at least one pair of rows labeled with different decisions. Sometimes, decision is omitted for *diagnostic* tables, where each row is assumed to have an unique decision.

A *test* for the table T is a subset of columns such that at the intersection with these columns any two rows with different decisions are different. The number of columns in a test is called its *length*. A *reduct* for T is a test for T for which each proper subset is not a test. It is clear that each test has a reduct as a subset.

1.2.2 Length of Tests

In this subsection, we consider results connected with the length of tests. Let $T = \{t_{ij}\}$ be a binary diagnostic decision table with m rows and n columns. We will say that columns $j_1,\ldots,j_k$ are *linearly dependent*, if $t_{ij_1} \oplus \ldots \oplus t_{ij_k} = 0$ for $i = 1,\ldots m$. We denote by *rank*(T) the maximal number of linearly independent columns in T. Soloviev in [68] noted that each reduct consists of linearly independent columns. On the other hand, a test must separate each pair of rows, so test length can not be less than a binary logarithm of the number of rows. This leads to the following theorem that gives an upper bound and a lower bound on the length of a reduct.

Theorem 1.47. *(Soloviev [68]) Let μ be length of some reduct for a diagnostic binary decision table T. Then $\lceil \log_2 m \rceil \leq \mu \leq \mathrm{rank}(T)$.*

A trivial upper bound on the length of reduct depending on the number of columns and rows is $\max(n, m-1)$; it can't be improved in general case. Preparata in [54] proved an upper bound on the average value of the minimum length of test that differs only by a multiplicative constant from the lower bound given by Theorem 1.47. This fact shows that for the most of tables, the minimum length of test is relatively close to its lower bound.

Theorem 1.48. *(Preparata [54]) Let $\bar{\mu}(n,m)$ be average minimum length of test, where the average is taken over all diagnostic binary decision tables with n columns and m rows. Then $\bar{\mu}(n,m) < \lceil 2\log_2 m \rceil$.*

Consider a diagnostic table T with m rows and n columns whose elements belong to the set $\{0,...,k-1\}$. Denote by $\mathscr{L}_k(m,n)$ the set of all such tables. We will say that certain property E *holds for almost all tables* from $\mathscr{L}_k(m,n)$ if $|\mathscr{L}_k^E(m,n)|/|\mathscr{L}_k(m,n)| \to 1$ with $n \to \infty$, where $\mathscr{L}_k^E(m,n)$ is the set of tables for which E holds. Recall that by $R(T)$ we denote the minimum length of a test for T.

The following theorem is proved by Korshunov in [31]. It gives a lower and an upper bound on the minimum length of a test that hold for almost all diagnostic tables.

Theorem 1.49. *(Korshunov [31])*

1. *If $m = m(n)$ satisfies the relation $2 \le m \le 1/3 \cdot \log_k n \cdot \log_2 \log_2 n$, then for almost all tables $T \in \mathscr{L}_k(m,n)$, $R(T) = \lceil \log_k m \rceil$.*
2. *If $m = m(n)$ satisfies the relation $1/3 \cdot \log_k n \cdot \log_2 \log_2 n < m \le k^{n/2}/\log_2 n$, then for almost all tables $T \in \mathscr{L}_k(m,n)$,*
$$\left\lfloor 2\log_k m - \log_k \left\{ nH\left(\frac{2\log_k m}{n}\right)\right\}\right\rfloor - 1 \le R(T) \le \left\lfloor 2\log_k m - \log_k \left\{ nH\left(\frac{2\log_k m}{n}\right)\right\}\right\rfloor + 5,$$
where $H(p) = -p\log_2 p - (1-p)\log_2(1-p)$.
3. *If $m = m(n)$ satisfies the relation $k^{n/2}/\log_2 n < m \le \sqrt{8k^{n-4}\log_2 n}$ and $\sqrt{k^{n-\nu}\nu\log_2 n} < m \le \sqrt{k^{n-\nu+1}(\nu-1)\log_2 n}$ for a natural ν, then for almost all tables $T \in \mathscr{L}_k(m,n)$, $n-\nu-2 \le R(T) \le n-\nu+1$.*
4. *If $m = m(n)$ satisfies the relation $\sqrt{8k^{n-4}\log_2 n} < m \le \sqrt{2k^n \log_2 n}$, then for almost all tables $T \in \mathscr{L}_k(m,n)$, $R(T) \ge n-3$.*
5. *If $m = m(n)$ satisfies the relation $\sqrt{2k^n\log_2 n} < m \le k^{n-1}$, then for almost all tables $T \in \mathscr{L}_k(m,n)$, $R(T) = n$.*
6. *If $m > k^{n-1}$, then for each table $T \in \mathscr{L}_k(m,n)$, $R(T) = n$.*

Noskov in [49] proved that in almost all binary diagnostic tables any $2\log_2 m$ columns form an asymptotically minimum test.

Theorem 1.50. *(Noskov [49]) If*
$$\lim_{m,n\to\infty} \frac{m}{2^{n/2}} = \lim_{m,n\to\infty} \frac{\log_2\log_2 n}{\log_2 m} = 0$$
then any fixed $2(1+\varepsilon)\log_2 m$ columns ($\varepsilon \to 0$ for $m \to \infty$) form a test in almost all tables from $\mathscr{L}_2(m,n)$.

Theorem 1.51. *(Noskov [49]) If $\lim_{m,n\to\infty} \log_2\log_2 n/\log_2 m = 0$ for $m, n \to \infty$, then almost all tables in $\mathscr{L}_2(m,n)$ do not have a test with length less than $2\log_2 m - \theta(m,n)$, where $\theta(m,n) \to \infty$, $\theta(m,n)/\log_2 m \to 0$ for $m,n \to \infty$.*

Additionally, a higher upper bound is proved for decision tables from $\mathscr{L}_2(m,n)$ without any restrictions on m and n.

Theorem 1.52. *(Noskov [49]) Almost all tables in* $\mathscr{L}_2(m,n)$ *do not have a reduct, whose length exceeds* $2\log_2 m + \log_2 n + 4$.

Zhuravlev in [76] considered *partially defined Boolean functions* that is another representation of binary decision tables. He proved an upper bound on the minimum length of test for almost all binary decision tables of a certain class. Denote by $\mathscr{M}_2(r,l,n)$ a class of binary decision tables with n columns, and $r+l$ rows, where r rows are labeled with the decision 0 and l rows with the decision 1.

Theorem 1.53. *(Zhuravlev [76]) If* $\lim_{n\to\infty} r(n)/2^{n/2} = \lim_{n\to\infty} l(n)/2^{n/2} = 0$, *then almost all tables from* $\mathscr{M}_2(r,l,n)$ *have a test of length* $2\log_2 \max(r,l) + \alpha(n)$, *where* $\lim_{n\to\infty} \alpha(n) = \infty$ *and* $\lim_{n\to\infty} \alpha(n)/\log_2 \max(r,l) = 0$.

Slepyan in [65] studied the minimum length of test in a class of diagnostic tables characterized by the number of zeros and ones. Let us choose a rational number p, $0 < p < 1$, and set into a correspondence to each table $T \in \mathscr{L}_2(m,n)$ the number $\omega(T) = p^{\nu} q^{mn-\nu}$, where $q = 1 - p$, and ν is the number of ones in the table T. Let us build a multiset $\tilde{\mathscr{L}}_2^p(m,n)$ by taking each table $T \in \mathscr{L}_2(m,n)$ the number of times proportional to $\omega(T)$.

Theorem 1.54. *(Slepyan [65]) For almost all tables in* $\tilde{\mathscr{L}}_2^p(m,n)$ *possessing the condition*

$$(\ln n)^{\mu} \leq m \leq (q^2 + p^2)^{-n/\rho(n)}, \mu > \frac{2\log(p^3+q^3)}{2\log(p^3+q^3)+3},$$

($\rho(n)$ is growing with n arbitrarily slowly), the minimum length of test is equal to k_0 *or* $k_0 + 1$, *where*

$$k_0 = \left\lceil \log \frac{m^2}{2s\ln(n/s)} - \frac{1}{2} \right\rceil,$$

$s = \log(m^2/(2\ln n))$, *and logarithms have a base* $(p^2+q^2)^{-1}$.

1.2.3 Number of Reducts

In this subsection, we consider results related mainly to the number of reducts.

A well-known upper bound on the maximal number of reducts for a binary diagnostic table is based on the result about the maximum number of independent binary tuples of the length n, that is $C_n^{\lfloor n/2 \rfloor}$. Soloviev in [68] proved a lower bound that differs from the upper bound by a multiplicative constant only.

Theorem 1.55. *(Soloviev [68]) Let* $Q(n)$ *be the maximal number of reducts for a diagnostic binary decision table with n columns. Then* $cC_n^{\lfloor n/2 \rfloor} < Q(n) < C_n^{\lfloor n/2 \rfloor}$, *where* $0.28 < c < 1$.

The lower bound can be generalized to the maximal number $Q_k(n)$ of reducts of a fixed length k for a diagnostic binary decision table with n columns. On the other hand, $Q_k(n)$ has a trivial upper bound C_n^k. This leads to the following statement.

Theorem 1.56. *(Soloviev [68])* $cC_n^k \le Q_k(n) < C_n^k$, *where* $c > 0.28$.

In [66] it is proved that upper bound is not reachable for any $n \ge 10$ and $4 \le k \le \lfloor n/2 \rfloor$

Theorem 1.57. *(Soloviev [66]) For any natural* n, $n \ge 10$,

$$Q_k(n) \le \begin{cases} C_n^k - \left\lfloor \frac{n-2k}{k+1} \right\rfloor (k+1), if\, 4 \le k < \left\lfloor \frac{n}{2} \right\rfloor, \\ C_n^k - k - 1, if\, k = \left\lfloor \frac{n}{2} \right\rfloor. \end{cases}$$

Under certain restriction on the number of rows, there is a similar result for $k > \lfloor n/2 \rfloor$.

Theorem 1.58. *(Soloviev [66]) For any natural* n *and* k, *such that* $\lfloor \frac{n}{2} \rfloor < k < n-1$ *there is no table with* n *columns and* m *rows that contains exactly* C_n^k *reducts of the length* k, *if*

$$m \le \frac{2^{(k-1)/2}}{\sqrt[4]{2\pi(n-k+1)}} \left(\frac{n}{k-1} \right)^{(2k-1)/4} e^{-1/12}.$$

Noskov in [49] gives a lower and an upper bound on the number of reducts for a certain class of diagnostic binary decision tables.

Theorem 1.59. *(Noskov [49]) If* $\log_2 \log_2 m / \log_2 n \to 0$ *and* $\log_2 \log_2 n / \log_2 m \to 0$ *for* $m, n \to \infty$ *then almost all tables from* $\mathcal{L}_2(m,n)$ *have more than* $n^{2(1-\delta)\log_2 m}$ *reducts, where* $\delta \to 0$ *for* $m, n \to \infty$.

Theorem 1.60. *(Noskov [49]) If* $\log_2 \log_2 m / \log_2 n \to 0$ *and* $\log_2 \log_2 n / \log_2 m \to 0$ *for* $m, n \to \infty$ *then almost all tables from* $\mathcal{L}_2(m,n)$ *have less than* $n^{2(1+\alpha)\log_2 m}$ *reducts, if* $2\log_2 m \ge \log_2 n$, *and less than* $n^{2(1+\beta)\log_2 n}$ *reducts, if* $2\log_2 m < \log_2 n$, *where* $\alpha, \beta \to 0$ *for* $m, n \to \infty$.

Additionally, a higher upper bound is proved for decision tables from $\mathcal{L}_2(m,n)$ without any restrictions on m and n.

Theorem 1.61. *(Noskov [49]) For almost all tables in* $\mathcal{L}_2(m,n)$ *the number of reducts is less than* $n^{2\log_2 m + \log_2 n + 5}$.

The paper [67] considers bounds on the number of diagnostic binary decision tables containing trivial reducts. For a table with m rows and n columns, a reduct is called *trivial* if its length ρ possesses the condition $\rho = \min(m-1, n)$.

We denote by $\Theta(m,n)$ the number of tables with m rows and n columns filled by numbers 0 and 1 that have at least one trivial reduct. Two cases are possible. The following theorem gives an upper and a lower bound on $\Theta(m,n)$, if $n \ge m-1$ (a trivial reduct has the length $m-1$ in this case).

Theorem 1.62. *(Soloviev [67]) Let* $n \ge m-1$ *and* C *be some positive constant. Then*

$$\frac{C}{m^2}(2e)^{(m-1)/3}\left(1 - e^{-m(m-1)/2^{n-m+2}}\right) \lambda(m,n) \le \Theta(m,n) \le 4^{m-1} \lambda(m,n),$$

where

$$\lambda(m,n) = \frac{2^{(m-1)}2^{(n-m+1)m}m!n!}{(n-m+1)!}.$$

If $m \geq n+1$, then a trivial reduct has the length n. The following theorem gives an upper and a lower bound on $\Theta(m,n)$ for this case.

Theorem 1.63. *(Soloviev [67]) Let $m \geq n+1$ and C be some positive constant. Then*

$$\Theta(m,n) \leq \begin{cases} n\frac{2^{n(n+4)}m!n!}{(m-2n)!}\left(2^n - n - \frac{m-1}{2}\right)^{m-2n}, \text{ if } m > 2n ; \\ (m-n)2^{4n}2^{(m-n)n}m!n!, \text{ if } n+1 \leq m \leq 2n ; \end{cases}$$

and

$$\Theta(m,n) \geq \begin{cases} C\left(\frac{2}{3}\right)^n \frac{2^{nq}(2e^2)^{n/3}\left((2^{n-1}-n)(2^{n-1}-m+q)\right)^{(m-n-q)/2}m!n!}{(m-n-q)!}, \\ \quad q = \lfloor\sqrt{2n}\rfloor, \text{ if } n+q < m < 2n ; \\ C\left(\frac{2}{3}\right)^n (2e^2)^{n/3}2^{(m-n)n}m!n!, \text{ if } n+1 \leq m \leq n+q . \end{cases}$$

We will say that two characteristics of decision tables A_{mn} and B_{mn} asymptotically coincide (and denote it $A_{mn} \sim B_{mn}$) if $\lim_{m,n\to\infty} A_{mn}/B_{mn} = 1$.

Noskov and Slepyan in [53] estimated the mean number of reducts for certain class of diagnostic binary decision tables and proved that for almost all tables from that class, the number of reducts asymptotically coincides to the mean value. Denote $\bar{Q}(m,n)$ the mean number of reducts for tables from $\mathcal{L}_2(m,n)$.

Theorem 1.64. *(Noskov and Slepyan[53])*

1. *If $2^{m^\beta} \geq n \geq m^\alpha$ $(\alpha > 2, \beta > 1/2)$, then $\bar{Q}(m,n) \sim \sum_{k\in\phi} C_n^k m^{2k}2^{-k-k^2}$ $(m,n \to \infty)$, where*

$$\phi = \left(\frac{1}{2}\log_2(nm^2) + \frac{1}{2}\log_2\log_2(nm^2) - \log_2\log_2\log_2 n ,\right.$$
$$\left.\frac{1}{2}\log_2(nm^2) + \frac{1}{2}\log_2\log_2(nm^2) + \log_2\log_2\log_2 n\right) .$$

2. *If $2^{m^\beta} \leq n \leq m^{-2}2^{2(m-2-\rho)}$ $(\rho = \lfloor\log_2(m\sqrt{n})\rfloor - \log_2(m\sqrt{n})))$, then $\log_2\bar{Q}(m,n) \sim 1/4\cdot\log_2^2 n$ $(m,n \to \infty)$.*
3. *If $n > m^{-2}2^{2(m-2-\rho)}$, then $\log_2\bar{Q}(m,n) \sim m(\log_2 n - m)$ $(m,n \to \infty)$.*

Theorem 1.65. *(Noskov and Slepyan [53]) If $2^{m^\beta} \geq n \geq m^\alpha$ $(\alpha > 2, \beta > 1/2)$, then for almost all tables from $\mathcal{L}_2(m,n)$ the number of reducts asymptotically coincides with $\bar{Q}(m,n)$.*

Slepyan in [64] considered average number of reducts for a certain class of diagnostic binary decision tables. He proved that there is a natural $k \leq \sqrt[3]{m}$ such that almost all reducts have length k or $k+1$. Denote $\mathcal{N}_2(m,n) \subset \mathcal{L}_2(m,n)$, a class of tables for which $n^\beta \leq m \leq 2^{\alpha n}$, $\alpha < \frac{1}{2}$, $\beta > \frac{1}{2}$. Denote by $\bar{Q}^*(m,n)$ and $\bar{Q}_k^*(m,n)$ average number of reducts and average number of reducts of the length k for tables from $\mathcal{N}_2(m,n)$.

Theorem 1.66. *(Slepyan [64])* $\bar{Q}^*(m,n) \sim \bar{Q}^*_k(m,n) + \bar{Q}^*_{k+1}(m,n)$ *for some* $k = k(m,n)$, $k \leq \sqrt[3]{m}$, $m,n \to \infty$.

For a table $T \in \mathcal{N}_2(m,n)$, denote the number of reducts by $Q(T)$. It was shown that the relative variance of the number of reducts in tables from $\mathcal{N}_2(m,n)$ tends to zero with growth of n.

Theorem 1.67. *(Slepyan [64]) For a randomly chosen table* $T \in \mathcal{N}_2(m,n)$,

$$P\left\{\left|\frac{Q(T)}{\bar{Q}^*(m,n)} - 1\right| < \varepsilon\right\} \to 1$$

for $n \to \infty$ *and an arbitrarily small positive* ε.

Additionally, an information weight of i-th column was studied that is fraction of reducts that contain i-th column. It was proved, that in tables from $\mathcal{N}_2(m,n)$, almost all columns have information weight close to $2/n \cdot \log_2 m$. For a table $T \in \mathcal{N}_2(m,n)$, denote the number of reducts containing i-th column by $Q^i(T)$, and denote $C^i(T) = Q^i(T)/Q(T)$ information weight of i-th column.

Theorem 1.68. *(Slepyan [64]) For a randomly chosen table* $T \in \mathcal{N}_2(m,n)$,

$$P\left\{\left|C^i(T) - \frac{2\log_2 m}{n}\right| < \varepsilon\right\} \to 1$$

for $n \to \infty$ *and an arbitrarily small positive* ε.

1.2.4 Algorithms for Construction of Minimum Tests

In this section we study exact and approximate algorithms for constructing a test, which has minimum length (see also Sect 1.1).

1.2.4.1 Exact Algorithm

An exact algorithm for building the set of reducts was first proposed by Chegis and Yablonskii in [10]. Let T be a decision table with n columns labeled with attributes $f_1,\dots,f_n$. Consider Boolean variables $x_1,\dots,x_n$. A subset of columns of T can be described by a vector of values of these variables as follows: for $i = 1,\dots,n$, x_i takes value 1 if the subset contains f_i, and 0 otherwise. Then we can describe a characteristic function of test by a Boolean formula as follows. For each pair of rows r_1, r_2 labeled with different decisions, we add to the formula an elementary disjunction $(x_{i_1} \vee \dots \vee x_{i_l})$ where $i_1,\dots,i_l$ are indices of columns, where r_1 and r_2 have different values. A product of elementary disjunctions for all pairs of rows labeled with different decisions gives a formula $F(T)$ that describes a function $f(x_1,\dots,x_n)$. One can see that f takes value 1 on a binary vector if and only if the corresponding subset of columns forms a test of the table T.

In order to get an explicit description of reducts, one should convert the formula $F(T)$ to the disjunctive normal form by multiplying elementary disjunctions and applying absorbtion of the kind $A \vee A\&B = A$ and $A\&A = A \vee A = A$ where it is possible. The resulted formula is a disjunction of elementary conjunctions of the kind $x_{j_1}\&\ldots\&x_{j_p}$. Each such conjunction describes a reduct, and all the set of conjunctions corresponds to the set of all reducts. Here, the conjunction $x_{j_1}\&\ldots\&x_{j_p}$ describes a reduct $\{f_{j_1},\ldots,f_{j_p}\}$.

Noskov in [49] gave an upper bound on complexity of the algorithm in terms of number of elementary multiplications (it's assumed that multiplication of two elementary disjunctions of length r and l requires rl elementary multiplications).

Theorem 1.69. *(Noskov [49]) For almost all tables from $\mathscr{L}_k(m,n)$, the algorithm for building of the set of reducts considered above requires at most $C_m^2 n^{4\log_2 m+2\log_2 n+12}$ elementary multiplications.*

1.2.4.2 Approximate Algorithm

Here we consider a greedy algorithm $\mathscr{A}$ that builds a test for a given decision table T. During each step, this algorithm chooses an attribute which separates the maximum number of pairs of rows with different decisions unseparated during the previous steps, and adds this attribute to the constructed test. The algorithm finishes when all pairs of rows with different decisions will be separated.

We denote by $R(T)$ the minimum length of a test for T, by $R_{\text{greedy}}(T)$ we denote the length of a test constructed by the algorithm $\mathscr{A}$, and by $P(T)$ – the number of unordered pairs of rows with different decisions in T.

Kospanov in [32] proved an upper bound on $R_{\text{greedy}}(T)$ for almost all diagnostic binary tables.

Theorem 1.70. *(Kospanov [32]) For almost all tables $T \in \mathscr{L}_2(m,n)$ possessing the condition $m^2/(2^{5/3-\log_2 3}n) \to 0$ for $n \to \infty$, $R_{\text{greedy}}(T) < c\log_2 m$, where c is some constant.*

From results obtained by Nigmatullin in [48] for greedy algorithm for set cover problem the following statement follows immediately

Theorem 1.71. *Let T be a decision table. Then*

$$R_{\text{greedy}}(T) \le R(T)(1+\ln P(T) - \ln R(T)) .$$

1.3 Infinite and Large Finite Sets of Attributes

In this section, we consider two approaches to the study of decision trees and decision rule systems for problems over infinite or large finite sets of attributes. Local approach is based on the assumption that only attributes contained in a problem description are used in decision trees and decision rules systems solving this problem. Global approach is based on the assumption that any attributes from the considered infinite or large finite set can be used in decision trees and decision rule systems solving the problem. This section is based mainly on monographs [42, 45].

The section contains three subsections. In Sect. 1.3.1, basic notions are discussed. Section 1.3.2 is devoted to the consideration of local approach for infinite and large finite sets of attributes. In Sect. 1.3.3, results related to the global approach are considered.

1.3.1 Basic Notions

Let A be a nonempty set, B be a finite nonempty set with at least two elements, and F be a nonempty set of functions from A to B. Functions from F will be called *attributes* and the triple $U = (A,B,F)$ will be called an *information system.* If F is a finite set then U will be called a *finite* information system. If F is an infinite set then U will be called an *infinite* information system.

We will consider problems over the information system U. A *problem over* U is an arbitrary $(n+1)$-tuple $z = (\nu, f_1, \ldots, f_n)$ where $\nu : B^n \to \omega$, $\omega = \{0,1,2,\ldots\}$, and $f_1,\ldots,f_n \in F$. The number $\dim z = n$ will be called the *dimension* of the problem z. The problem z may be interpreted as a problem of searching for the value $z(a) = \nu(f_1(a),\ldots,f_n(a))$ for an arbitrary $a \in A$. Different problems of pattern recognition, discrete optimization, fault diagnosis and computational geometry can be represented in such form. We denote by $\mathscr{P}(U)$ the set of all problems over the information system U.

As algorithms for problem solving we will consider decision trees and decision rule systems.

A *decision tree over* U is a marked finite tree with the root in which each terminal node is labeled with a number from ω; each node which is not terminal (such nodes are called *working*) is labeled with an attribute from F; each edge is labeled with an element from B. Edges starting in a working node are labeled with pairwise different elements.

Let Γ be a decision tree over U. A *complete path* in Γ is an arbitrary sequence $\xi = v_1, d_1, \ldots, v_m, d_m, v_{m+1}$ of nodes and edges of Γ such that v_1 is the root, v_{m+1} is a terminal node, and v_i is the initial and v_{i+1} is the terminal node of the edge d_i for $i = 1,\ldots,m$. Now we define a system of equations $\mathscr{S}(\xi)$ and a subset $\mathscr{A}(\xi)$ of the set A associated with ξ. If $m = 0$ then $\mathscr{S}(\xi)$ is empty system and $\mathscr{A}(\xi) = A$. Let $m > 0$, the node v_i be labeled with the attribute f_i, and the edge d_i be labeled with the element δ_i from B, $i = 1,\ldots,m$. Then $\mathscr{S}(\xi) = \{f_1(x) = \delta_1, \ldots, f_m(x) = \delta_m\}$ and $\mathscr{A}(\xi)$ is the set of solutions of $\mathscr{S}(\xi)$ from A.

We will say that a decision tree Γ over U *solves* a problem z over U if for any $a \in A$ there exists a complete path ξ in Γ such that $a \in \mathscr{A}(\xi)$, and the terminal node of the path ξ is labeled with the number $z(a)$.

For decision trees, as time complexity measure we will consider the *depth* of a decision tree which is the maximum number of working nodes in a complete path in the tree. As space complexity measure we will consider the number of nodes in a decision tree. We denote by $h(\Gamma)$ the depth of a decision tree Γ, and by $\#(\Gamma)$ we denote the number of nodes in Γ. Note that for each problem z over U there exists a decision tree Γ over U which solves the problem z and for which $h(\Gamma) \le \dim z$ and $\#(\Gamma) \le |B|^{\dim z+1}$.

A *decision rule over* U is an arbitrary expression of the kind

$$f_1 = \delta_1 \wedge \ldots \wedge f_m = \delta_m \rightarrow \sigma$$

where $f_1, \ldots, f_m \in F$, $\delta_1, \ldots, \delta_m \in B$ and $\sigma \in \omega$. We denote this decision rule by ρ. The number m will be called the *length* of the rule ρ. Now we define a system of equations $\mathscr{S}(\rho)$ and a subset $\mathscr{A}(\rho)$ of the set A associated with ρ. If $m = 0$ then $\mathscr{S}(\rho)$ is empty system and $\mathscr{A}(\rho) = A$. Let $m > 0$. Then $\mathscr{S}(\rho) = \{f_1(x) = \delta_1, \ldots, f_m(x) = \delta_m\}$ and $\mathscr{A}(\rho)$ is the set of solutions of $\mathscr{S}(\rho)$ from A. The number σ will be called the *right-hand side of the rule* ρ.

A *decision rule system over* U is a nonempty finite set of decision rules over U. Let S be a decision rule system over U and z be a problem over U. We will say that the decision rule system S is *complete for the problem* z if for any $a \in A$ there exists a rule $\rho \in S$ such that $a \in \mathscr{A}(\rho)$, and for each rule $\rho \in S$ such that a is a solution of $\mathscr{S}(\rho)$, the right-hand side of ρ coincides with the number $z(a)$.

For decision rule systems, as time complexity measure we will consider the maximum length $L(S)$ of a rule from the system S. We will say about $L(S)$ as about the *depth* of decision rule system S. As space complexity measure we will consider the number of rules in a system. Note that for each problem z over U there exists a decision rule system S over U which is complete for the problem z and for which $L(S) \leq \dim z$ and $|S| \leq |B|^{\dim z}$.

The investigation of decision trees and decision rule systems for a problem $z = (\nu, f_1, \ldots, f_n)$ which use only attributes from the set $\{f_1, \ldots, f_n\}$ is based on the study of the *decision table* $T(z)$ associated with the problem z. The table $T(z)$ is a rectangular table with n columns which contains elements from B. The row $(\delta_1, \ldots, \delta_n)$ is contained in the table $T(z)$ if and only if the equation system

$$\{f_1(x) = \delta_1, \ldots, f_n(x) = \delta_n\}$$

is compatible on A (has a solution from the set A). This row is labeled with the number $\nu(\delta_1, \ldots, \delta_n)$. For $i = 1, \ldots, n$, the i-th column is labeled with the attribute f_i. We know that a decision tree over $T(z)$ solves the problem z if and only if this tree is a decision tree for $T(z)$. We know also that a decision rule system over $T(z)$ is complete for the problem z if and only if this system is a complete decision rule system for $T(z)$.

If we would like to consider additional attributes $f_{n+1}, \ldots, f_{n+m}$, we can study a new problem $z' = (\mu, f_1, \ldots, f_n, f_{n+1}, \ldots, f_{n+m})$ such that

$$\mu(x_1, \ldots, x_{n+m}) = \nu(x_1, \ldots, x_n) ,$$

and corresponding decision table $T(z')$.

1.3.2 Local Approach to Study of Decision Trees and Rules

Let $U = (A, B, F)$ be an information system. For a problem $z = (\nu, f_1, \ldots, f_n)$ over U we denote by $h_U^l(z)$ the minimum depth of a decision tree over U which solves the problem z and uses only attributes from the set $\{f_1,\ldots,f_n\}$. We denote by $L^l_U(z)$ the minimum depth of a complete decision rule system for z which uses only attributes from the set $\{f_1,\ldots,f_n\}$. We will consider relationships among the parameters $h^l_U(z)$, $L^l_U(z)$ and $\dim z$. To this end, we define the functions $h^l_U : \omega\setminus\{0\}\to\omega$ and $L^l_U : \omega\setminus\{0\}\to\omega$ in the following way:

$$h^l_U(n) = \max\{h^l_U(z) : z\in\mathscr{P}(U), \dim z\le n\}\,,$$
$$L^l_U(n) = \max\{L^l_U(z) : z\in\mathscr{P}(U), \dim z\le n\}$$

for any $n\in\omega\setminus\{0\}$, where $\mathscr{P}(U)$ is the set of all problems over U. The functions h^l_U and L^l_U are called the *local Shannon functions* for the information system U. Using Corollary 1.6 we obtain $L^l_U(n)\le h^l_U(n)$ for any $n\in\omega\setminus\{0\}$.

1.3.2.1 Arbitrary Information Systems

We will say that the information system $U=(A,B,F)$ satisfies the *condition of reduction* if there exists a number $m\in\omega\setminus\{0\}$ such that for each compatible on A system of equations

$$\{f_1(x)=\delta_1,\ldots,f_r(x)=\delta_r\}$$

where $r\in\omega\setminus\{0\}$, $f_1,\ldots,f_r\in F$ and $\delta_1,\ldots,\delta_r\in B$ there exists a subsystem of this system which has the same set of solutions and contains at most m equations.

In the following theorem, the criterions of the local Shannon function h^l_U behavior are considered.

Theorem 1.72. *Let U be an information system. Then the following statements hold:*

a) if U is a finite information system then $h^l_U(n)=O(1)$;

b) if U is an infinite information system which satisfies the condition of reduction then $h^l_U(n)=\Theta(\log n)$;

c) if U is an infinite information system which does not satisfy the condition of reduction then $h^l_U(n)=n$ for each $n\in\omega\setminus\{0\}$.

There are only two types of behavior of the local Shannon function L^l_U.

Theorem 1.73. *Let $U=(A,B,F)$ be an information system. Then the following statements hold:*

a) if U is a finite information system then $L^l_U(n)=O(1)$;

b) if U is an infinite information system which satisfies the condition of reduction then $L^l_U(n)=O(1)$;

c) if U is an infinite information system which does not satisfy the condition of reduction then $L^l_U(n)=n$ for each $n\in\omega\setminus\{0\}$.

Now we consider an example.

Example 1.74. Let $m,t\in\omega\setminus\{0\}$. We denote by $Pol(m)$ the set of all polynomials which have integer coefficients and depend on variables $x_1,\ldots,x_m$. We denote by

$Pol(m,t)$ the set of all polynomials from $Pol(m)$ such that the degree of each polynomial is at most t. We define information systems $U(m)$ and $U(m,t)$ as follows: $U(m) = (\mathbb{R}^m, E, F(m))$ and $U(m,t) = (\mathbb{R}^m, E, F(m,t))$ where $E = \{-1, 0, +1\}$, $F(m) = \{\text{sign}(p) : p \in Pol(m)\}$ and $F(m,t) = \{\text{sign}(p) : p \in Pol(m,t)\}$. Here $\text{sign}(x) = -1$ if $x < 0$, $\text{sign}(x) = 0$ if $x = 0$, and $\text{sign}(x) = +1$ if $x > 0$. One can prove that $h^l_{U(m)}(n) = L^l_{U(m)}(n) = n$ for each $n \in \omega \setminus \{0\}$, $h^l_{U(1,1)}(n) = \Theta(\log_2 n)$, $L^l_{U(1,1)}(n) = O(1)$, and if $m > 1$ or $t > 1$ then $h^l_{U(m,t)}(n) = L^l_{U(m,t)}(n) = n$ for each $n \in \omega \setminus \{0\}$.

1.3.2.2 Restricted Information Systems

An information system is called *restricted* if it satisfies the condition of reduction. Let $U = (A, B, F)$ be a restricted information system. Then there exists a number $m \in \omega \setminus \{0\}$ such that for each compatible on A system of equations

$$\{f_1(x) = \delta_1, \dots, f_r(x) = \delta_r\},$$

where $r \in \omega \setminus \{0\}$, $f_1, \dots, f_r \in F$ and $\delta_1, \dots, \delta_r \in B$, there exists a subsystem of this system which has the same set of solutions and contains at most m equations.

Let $z = (\nu, f_1, \dots, f_n)$ be a problem over U and $T(z)$ be the decision table corresponding to this problem. It was shown in [43] that

$$|SEP(T(z))| \leq |B|^m n^m + 1,$$

where $SEP(T(z))$ is the set of all separable subtables of the table $T(z)$ including $T(z)$. From this inequality and from Theorems 1.44 and 1.45 it follows that the algorithms W and V for exact optimization of decision trees and rules have on tables $T(z)$ polynomial time complexity depending on $\dim z$.

In [43] it was shown that if U is not a restricted information system then, for any natural n, there exists a problem z over U such that $\dim z \leq n$ and

$$|SEP(T(z))| \geq 2^n.$$

The mentioned facts determine the interest of studying restricted information systems.

We now consider restricted binary linear information systems in the plane. Let P be the set of all points in the plane and l be a straight line (line in short) in the plane. This line divides the plane into two open half-planes H_1 and H_2 and the line l. Two attributes *correspond* to the line l. The first attribute takes value 0 on points from H_1, and value 1 on points from H_2 and l. The second one takes value 0 on points from H_2, and value 1 on points from H_1 and l. We denote by $\mathcal{L}$ the set of all attributes corresponding to lines in the plane. Information systems of the kind $(P, \{0,1\}, F)$, where $F \subseteq \mathcal{L}$, will be called *binary linear information systems in the plane*. We will describe all restricted binary linear information systems in the plane.

Let l be a line in the plane. We denote by $\mathcal{L}(l)$ the set of all attributes corresponding to lines which are parallel to l. Let p be a point in the plane. We denote by

$\mathscr{L}(p)$ the set of all attributes corresponding to lines which pass through p. A set C of attributes from $\mathscr{L}$ will be called a *clone* if $C \subseteq \mathscr{L}(l)$ for some line l or $C \subseteq \mathscr{L}(p)$ for some point p.

Theorem 1.75. *A binary linear information system in the plane $(P,\{0,1\},F)$ is restricted if and only if F is the union of a finite number of clones.*

1.3.2.3 Local Shannon Functions for Finite Information Systems

Theorems 1.72 and 1.73 give us some information about the behavior of local Shannon functions for infinite information systems. But for a finite information system U we have only the relations $h_U^l(n) = O(1)$ and $L_U^l(n) = O(1)$. However, finite information systems are important for different applications.

Now we consider the behavior of the local Shannon functions for an arbitrary finite information system $U = (A,B,F)$ such that $f \not\equiv \text{const}$ for any $f \in F$.

A set $\{f_1,\dots,f_n\} \subseteq F$ will be called *redundant* if $n \geq 2$ and there exist $i \in \{1,\dots,n\}$ and $\mu : B^{n-1} \to B$ such that

$$f_i(a) = \mu(f_1(a),\dots,f_{i-1}(a),f_{i+1}(a),\dots,f_n(a))$$

for each $a \in A$. If the set $\{f_1,\dots,f_n\}$ is not redundant then it will be called *irredundant*. We denote by $\text{ir}(U)$ the maximum number of attributes in an irredundant subset of the set F.

A *systems of equations over U* is an arbitrary system

$$\{f_1(x) = \delta_1,\dots,f_n(x) = \delta_n\} \tag{1.1}$$

such that $n \in \omega \setminus \{0\}$, $f_1,\dots,f_n \in F$ and $\delta_1,\dots,\delta_n \in B$. The system (1.1) will be called *cancelable* if $n \geq 2$ and there exists a number $i \in \{1,\dots,n\}$ such that the system

$$\{f_1(x) = \delta_1,\dots,f_{i-1}(x) = \delta_{i-1},f_{i+1}(x) = \delta_{i+1},\dots,f_n(x) = \delta_n\}$$

has the same set of solutions just as the system (1.1). If the system (1.1) is not cancelable then it will be called *uncancelable*. We denote by $\text{un}(U)$ the maximum number of equations in an uncancelable compatible system over U.

One can show that

$$1 \leq \text{un}(U) \leq \text{ir}(U)\,.$$

The values $\text{un}(U)$ and $\text{ir}(U)$ will be called *the first and the second local critical points of the information system $U = (A,B,F)$*. Now we describe the behaviors of the local Shannon functions in terms of local critical points of U and the cardinality of the set B.

Theorem 1.76. *Let $U = (A,B,F)$ be a finite information system such that $f \not\equiv \text{const}$ for any $f \in F$, and $n \in \omega \setminus \{0\}$. Then the following statements hold:*

a) if $n \leq \text{un}(U)$ then $h_U^l(n) = n$;

b) if $\text{un}(U) \leq n \leq \text{ir}(U)$ *then*

$$\max\{\text{un}(U), \log_k(n+1)\} \leq h_U^l(n) \leq \min\{n, 2(\text{un}(U))^2 \log_2 2(kn+1)\}$$

where $k = |B|$;
c) if $n \geq \text{ir}(U)$ *then* $h_U^l(n) = h_U^l(\text{ir}(U))$.

Theorem 1.77. *Let* $U = (A, B, F)$ *be a finite information system such that* $f \not\equiv \text{const}$ *for any* $f \in F$, *and* $n \in \omega \setminus \{0\}$. *Then the following statements hold:*

a) if $n \leq \text{un}(U)$ *then* $L_U^l(n) = n$;
b) if $n \geq \text{un}(U)$ *then* $L_U^l(n) = \text{un}(U)$.

Of course, the problem of computing the values $\text{un}(U)$ and $\text{ir}(U)$ for a given finite information system U is very complicated problem. But obtained results allow us to constrict essentially the class of possible types of local Shannon functions for finite information systems.

Example 1.78. We denote by P the set of all points in the plane. Let us consider an arbitrary straight line l, which divides the plane into positive and negative open half-planes, and the line l itself. We assign a function $f : P \to \{0, 1\}$ to the line l. The function f takes the value 1 if a point is situated on the positive half-plane, and f takes the value 0 if a point is situated on the negative half-plane or on the line l. We denote by F the set of functions which correspond to certain r mutually disjoint finite classes of parallel straight lines. Let us consider a finite information system $U = (P, \{0, 1\}, F)$. One can show that $\text{ir}(U) = |F|$ and $\text{un}(U) \leq 2r$.

1.3.3 Global Approach to Study of Decision Trees and Rules

First, we consider arbitrary infinite information systems. Later, we will study two-valued finite information systems.

1.3.3.1 Infinite Information Systems

Let $U = (A, B, F)$ be an infinite information system.

We now define the notion of *independence dimension* (or, in short, *I-dimension*) of information system U. A finite subset $\{f_1, \ldots, f_p\}$ of the set F is called an *independent set* if there exist two-element subsets $B_1, \ldots, B_p$ of the set B such that for any $\delta_1 \in B_1, \ldots, \delta_p \in B_p$ the system of equations

$$\{f_1(x) = \delta_1, \ldots, f_p(x) = \delta_p\} \tag{1.2}$$

is compatible on the set A (has a solution from A). If for any natural p there exists a subset of the set F, which cardinality is equal to p and which is an independent set, then we will say that the information system U has infinite I-dimension. Otherwise, I-dimension of U is the maximum cardinality of a subset of F, which is an independent set.

The notion of I-dimension is closely connected with well known notion of Vapnik-Chervonenkis dimension [70]. In particular, an information system $(A, \{0,1\}, F)$ has finite I-dimension if and only if it has finite VC-dimension [34].

Now we consider the condition of decomposition for the information system U. Let $p \in \omega \setminus \{0\}$. A nonempty subset D of the set A will be called (p,U)-set if D coincides with the set of solutions on A of a system of the kind (1.2) where $f_1, \ldots, f_p \in F$ and $\delta_1, \ldots, \delta_p \in B$ (we admit that among the attributes $f_1, \ldots, f_p$ there are identical ones).

We will say that the information system U satisfies the *condition of decomposition* if there exist numbers $m, t \in \omega \setminus \{0\}$ such that every $(m+1,U)$-set is a union of t sets each of which is an (m,U)-set (we admit that among the considered t sets there are identical ones).

We consider partition of the set of infinite information systems into two classes: $\mathscr{C}_1$ and $\mathscr{C}_2$. The class $\mathscr{C}_1$ consists of all infinite information systems each of which has finite I-dimension and satisfies the condition of decomposition. The class $\mathscr{C}_2$ consists of all infinite information systems each of which has infinite I-dimension or does not satisfy the condition of decomposition.

In the following theorem, time and space complexity of decision trees are considered.

Theorem 1.79. *Let $U = (A,B,F)$ be an infinite information system. Then the following statements hold:*

a) if $U \in \mathscr{C}_1$ then for any ε, $0 < \varepsilon < 1$, there exists a positive constant c such that for each problem z over U there exists a decision tree Γ over U which solves the problem z and for which $h(\Gamma) \leq c(\log_2 n)^{1+\varepsilon} + 1$ and $\#(\Gamma) \leq |B|^{c(\log_2 n)^{1+\varepsilon}+2}$ where $n = \dim z$;

b) if $U \in \mathscr{C}_1$ then for any $n \in \omega \setminus \{0\}$ there exists a problem z over U with $\dim z = n$ such that for each decision tree Γ over U, which solves the problem z, the inequality $h(\Gamma) \geq \log_{|B|}(n+1)$ holds;

c) if $U \in \mathscr{C}_2$ then for any $n \in \omega \setminus \{0\}$ there exists a problem z over U with $\dim z = n$ such that for each decision tree Γ over U, which solves the problem z, the inequality $h(\Gamma) \geq n$ holds.

In the next theorem, time and space complexity of decision rule systems are considered.

Theorem 1.80. *Let $U = (A,B,F)$ be an infinite information system. Then the following statements hold:*

a) if $U \in \mathscr{C}_1$ then for any ε, $0 < \varepsilon < 1$, there exist positive constants c_1 and c_2 such that for each problem z over U there exists a decision rule system S over U which is complete for the problem z and for which $L(S) \leq c_1$ and $|S| \leq |B|^{c_2(\log_2 n)^{1+\varepsilon}+1}$ where $n = \dim z$;

b) if $U \in \mathscr{C}_2$ then for any $n \in \omega \setminus \{0\}$ there exists a problem z over U with $\dim z = n$ such that for each decision rule system S over U, which is complete for the problem z, the inequality $L(S) \geq n$ holds or the inequality $|S| \geq 2^n$ holds.

So the class $\mathscr{C}_1$ is interesting from the point of view of different applications. The following example characterizes both the wealth and the boundedness of this class.

Example 1.81. Let $m,t \in \omega \setminus \{0\}$. We consider the same information systems $U(m)$ and $U(m,t)$ as in Example 1.74. One can prove that $U(m) \in \mathscr{C}_2$ and $U(m,t) \in \mathscr{C}_1$. Note that the system $U(m)$ has infinite I-dimension.

1.3.3.2 Quasilinear Information Systems

We consider here bounds on complexity of decision trees and decision rule systems over quasilinear information systems. Such systems are important in applications of TT to discrete optimization (see Sect. 1.6).

We will call a set K a *numerical ring with unity* if $K \subseteq \mathbb{R}$, $1 \in K$, and for every $a,b \in K$ the relations $a+b \in K$, $a \times b \in K$ and $-a \in K$ hold. For instance, $\mathbb{R}$, $\mathbb{Q}$ and $\mathbb{Z}$ are numerical rings with unity.

Let K be a numerical ring with unity, A be a nonempty set and let $\varphi_1,\dots,\varphi_m$ be functions from A to $\mathbb{R}$. We denote

$$F(A,K,\varphi_1,\dots,\varphi_m) = \left\{ \operatorname{sign}\left(\sum_{i=1}^{m} d_i\varphi_i(x) + d_{m+1} \right) : d_1,\dots,d_{m+1} \in K \right\} .$$

Here $\operatorname{sign}(x) = -1$ if $x < 0$, $\operatorname{sign}(x) = 0$ if $x = 0$, and $\operatorname{sign}(x) = +1$ if $x > 0$. Set $E = \{-1,0,+1\}$. The information system $(A,E,F(A,K,\varphi_1,\dots,\varphi_m))$ will be denoted by $U(A,K,\varphi_1,\dots,\varphi_m)$ and will be called a *quasilinear information system.*

Let $f \in F(A,K,\varphi_1,\dots,\varphi_m)$ and $f = \operatorname{sign}(\sum_{i=1}^{m} d_i\varphi_i(x) + d_{m+1})$. We define the parameter $r(f)$ of the attribute f as follows. If $(d_1,\dots,d_{m+1}) = (0,\dots,0)$ then $r(f) = 0$. Otherwise,

$$r(f) = \max\{0, \max\{\log_2 |d_i| : i \in \{1,\dots,m+1\}, d_i \neq 0\}\} .$$

For a problem $z = (\nu, f_1,\dots,f_n)$ over $U(A,K,\varphi_1,\dots,\varphi_m)$, set $r(z) = \max\{r(f_1), \dots, r(f_n)\}$. Let Γ be a decision tree over $U(A,K,\varphi_1,\dots,\varphi_m)$ and $F(\Gamma)$ be the set of all attributes attached to nonterminal nodes of Γ. We denote $r(\Gamma) = \max\{r(f) : f \in F(\Gamma)\}$ (if $F(\Gamma) = \emptyset$ then $r(\Gamma) = 0$).

Theorem 1.82. *Let $U = U(A,K,\varphi_1,\dots,\varphi_m)$ be a quasilinear information system. Then for each problem z over U there exists a decision tree Γ over U which solves the problem z and for which the following inequalities hold:*

$$h(\Gamma) \leq (2(m+2)^3 \log_2(\dim z + 2m + 2))/\log_2(m+2) ,$$

$$r(\Gamma) \leq 2(m+1)^2(r(z) + 1 + \log_2(m+1)) .$$

Let $U = U(A,K,\varphi_1,\dots,\varphi_m)$ be a quasilinear information system. For a problem $z = (\nu, f_1,\dots,f_n)$ over U, we denote by $h_U^g(z)$ the minimum depth of a decision tree over U which solves the problem z. By $L_U^g(z)$ we denote the minimum depth of a decision rule system over U which is complete for the problem z. We define two functions $h_U^g : \omega \setminus \{0\} \to \omega$ and $L_U^g : \omega \setminus \{0\} \to \omega$ in the following way:

$$h_U^g(n) = \max\{h_U^g(z) : z \in \mathscr{P}(U), \dim z \leq n\} ,$$
$$L_U^g(n) = \max\{L_U^g(z) : z \in \mathscr{P}(U), \dim z \leq n\}$$

for any $n \in \omega \setminus \{0\}$. The functions h_U^g and L_U^g are called *global Shannon functions* for the information system U.

The following theorem is a simple corollary of Theorems 1.79, 1.80 and 1.82.

Theorem 1.83. *Let $U = U(A, K, \varphi_1, \ldots, \varphi_m)$ be a quasilinear information system. Then the following statements hold:*

a) if $\{(\varphi_1(a), \ldots, \varphi_m(a)) : a \in A\}$ is a finite set then $h_U^g(n) = O(1)$;
b) if $\{(\varphi_1(a), \ldots, \varphi_m(a)) : a \in A\}$ is an infinite set then $h_U^g(n) = \Theta(\log_2 n)$;
c) $L_U^g(n) = O(1)$.

1.3.3.3 Global Shannon Function h_U^g for Two-Valued Finite Information Systems

An information system $U = (A, B, F)$ will be called *two-valued* if $|B| = 2$. For a problem $z = (\nu, f_1, \ldots, f_n)$ over U, we denote by $h_U^g(z)$ the minimum depth of a decision tree over U which solves the problem z. We define the function $h_U^g : \omega \setminus \{0\} \to \omega$ in the following way:

$$h_U^g(n) = \max\{h_U^g(z) : z \in \mathscr{P}(U), \dim z \leq n\}$$

for any $n \in \omega \setminus \{0\}$. The function h_U^g will be called a *global Shannon function* for the information system U.

Now we consider the behavior of this global Shannon function for an arbitrary two-valued finite information system $U = (A, B, F)$ such that $f \not\equiv \text{const}$ for any $f \in F$.

Recall that by $\text{ir}(U)$ we denote the maximum number of attributes in an irredundant subset of the set F (see Sect. 1.3.2.3).

A problem $z \in P(U)$ will be called *stable* if $h_U^g(z) = \dim z$. We denote by $\text{st}(U)$ the maximum dimension of a stable problem over U.

One can show that

$$1 \leq \text{st}(U) \leq \text{ir}(U) .$$

The values $\text{st}(U)$ and $\text{ir}(U)$ will be called *the first and the second global critical points of the information system U*. Now we describe the behavior of the global Shannon function h_U^g in terms of global critical points of U.

Theorem 1.84. *Let U be a two-valued finite information system such that $f \not\equiv \text{const}$ for any $f \in F$, and $n \in \omega \setminus \{0\}$. Then the following statements hold:*

a) if $n \leq \text{st}(U)$ then $h_U^g(n) = n$;
b) if $\text{st}(U) < n \leq \text{ir}(U)$ then

$$\max\{\text{st}(U), \log_2(n+1)\} \leq h_U^g(n) \leq \min\left\{n, 8(\text{st}(U)+1)^5(\log_2 n)^2\right\} ;$$

c) if $n \geq \text{ir}(U)$ then $h_U^g(n) = h_U^g(\text{ir}(U))$.

The problem of computing the values $\mathrm{st}(U)$ and $\mathrm{ir}(U)$ for a given two-valued finite information system U is a complicated problem. However, the obtained results allow us to constrict the class of possible types of the global Shannon function h_U^g.

Example 1.85. Let us consider the same information system $U = (P, \{0,1\}, F)$ as in Example 1.78. One can show that $\mathrm{st}(U) \leq 2r$ and $\mathrm{ir}(U) = |F|$.

1.4 Pattern Recognition

In this section, we consider very briefly TT approaches to resolve problems of pattern recognition. They can be formulated as problems of prediction. Let us assume that we have a partial information about decision table T' represented in the form of a decision table T with the same set of attributes. The table T is a part of the table T'. Based on T we can construct a classifier which will predict values of the decision attribute for rows of T' using values of conditional attributes.

At the beginning, TT was oriented on the study of fault control and diagnosis problems. For these problems, the corresponding decision tables are large but (on principle) known. The first publication related to the problem of prediction was the paper of Dmitriev, Zhuravlev, and Krendelev [16]. It was inspired by complicated problems in geology connected with recognition of special type of gold deposits. In this paper, it was proposed, in particular, to measure the importance of an attribute by the number of reducts of T to which this attribute belongs. Later, many problems in geology, medicine, economics, etc., were resolved by recognition methods based on TT [17, 30, 77].

There are different approaches to design classifiers for the decision table T. One of the simplest ways is to construct a decision tree for the table T. We can apply this tree to a row of T'. If the computation is finished in a terminal node then the number attached to this node will be predicted value of the decision attribute.

Another way (known beginning with the first publications about "Kora" type algorithms [6, 69]) is to design a decision rule system for the table T and to use it to predict values of the decision attribute based on values of the conditional attributes. To this end we can apply various procedures of voting. In particular, any rule with left-hand side true for a row of T' can be considered as a vote "pro" the decision from the right-hand side of this rule.

We can use tests for T to predict values of the decision attribute for rows of T'. Let $\{f_{i_1}, \ldots, f_{i_m}\}$ be a test for T. For each row r of T, we construct the decision rule

$$f_{i_1} = b_1 \wedge \ldots \wedge f_{i_m} = b_m \rightarrow t$$

where $b_1, \ldots, b_m$ are numbers at the intersection of the row r and columns $f_{i_1}, \ldots, f_{i_m}$, and t is the decision attached to the row r. As a result we will obtain a system of true for T decision rules that can be used as a classifier.

Another approach is to use an ensemble of classifiers instead of one classifier (see survey of Kudriavtsev and Andreev [33] for details). We choose some set F of tests for the decision table T and for each test from F we construct a classifier as it was described above. We apply each classifier to a given row of the decision table T' and after that use a procedure of voting to choose a value of the decision attribute for the considered row. As the set F we can choose the set of all tests for T, the set of all reducts for T, the set of all tests with minimum length (cardinality) for T, the set of short tests which length is "close" to logarithm of the number of conditional attributes in T, etc.

The problem of the set F construction is usually very complicated. In particular, the number of reducts grows exponentially in the worst case with the growth of decision table size. Even if we consider the number of reducts as a parameter of our algorithm, the problem of construction of the set of reducts in polynomial time depending on the size of table and number of reducts, continue to be nontrivial.

We consider here one of approaches to the set of reduct construction proposed by Dyukova (see a survey of Dyukova and Zhuravlev [19] for details). Let T contain n columns labeled with conditional attributes $f_1,\ldots,f_n$. We will correspond to T a 0-1-matrix L_T with n columns. Each row r of L_T is obtained from a pair of rows r_1,r_2 of T labeled with different decisions. At the intersection with the column j, $j\in\{1,\ldots,n\}$, the row r has 1 if and only if the values of attribute f_j for the rows r_1 and r_2 are different. One can show that a subset $\{f_{i_1},\ldots,f_{i_m}\}$ of the set $\{f_1,\ldots,f_n\}$ is a reduct for T if and only if the following conditions hold: (i) each row of L_T has at least one 1 at the intersection with columns $i_1,\ldots,i_m$; (ii) for any $j\in\{1,\ldots,m\}$, there is a row of L_T which has only one 1 at the intersection with columns $i_1,\ldots,i_m$ and this 1 is at the intersection with the column i_j. So to find all reducts $\{f_{i_1},\ldots,f_{i_m}\}$ for the table T we should find all subsets $\{i_1,\ldots,i_m\}$ of the set $\{1,\ldots,n\}$ for which the set of columns of L_T with numbers $i_1,\ldots,i_m$ satisfies conditions (i) and (ii).

Let us consider 0-1-matrices with n columns and u rows where $\log_2 n\le u\le n^{1-\varepsilon}$, $\varepsilon>0$. It was shown that for almost all such matrices, the number of subsets of columns satisfying the condition (ii) is asymptotically equal (when $n\to\infty$) to the number of subsets of columns satisfying both conditions (i) and (ii). So instead of construction of subsets satisfying conditions (i) and (ii) we can construct subsets satisfying only the condition (ii) which is essentially easier problem.

Andreev created algorithms for construction of the sets of reducts for decision tables with number of rows greater than the number of attributes. Kibkalo designed algorithms for construction of the set of short tests which can be more stable than the set of reducts or the set of minimum tests under table changes. There are some papers which are devoted to the study of stochastic algorithms for construction of the set of reducts or for the evaluation of the number of reducts (details can be found in [33]).

1.5 Control and Diagnosis of Faults

One of the most developed areas of TT applications is control and diagnosis of faults. The first papers in this area were published in fifties and sixties of the last century by Yablonskii, Chegis, Moore, Eldred, Armstrong, Roth [5, 10, 20, 37, 59, 75]. We can not consider all results obtained here, in particular, we will not consider results for automata (see papers of Yablonskii and Vasilevskii [71, 72, 73, 74]). We will concentrate on results for contact networks and combinatorial circuits. Note that a part of this section is based on survey of Yablonskii [74].

This section consists of five subsections. In Sects. 1.5.1 and 1.5.2, main notions are discussed. Section 1.5.3 is devoted to the study of faults on inputs of nets. In Sect. 1.5.4, we consider problems of control and diagnosis for arbitrary (not specially constructed) combinatorial circuits from some classes. In Sect. 1.5.5, we study specially constructed nets.

1.5.1 Problems of Control and Diagnosis

Let S be either contact network or combinatorial circuit which implements a Boolean function $f(x_1,\dots,x_n)$. Let I be a source of faults that transforms S into one of nets from finite set $I(S)$. We assume for simplicity that $S \in I(S)$. Let $f_1(x_1,\dots,x_n),\dots, f_m(x_1,\dots,x_n)$ be all functions implemented by nets from $I(S)$ and $f_1 = f$.

The *problem of control*: for a given net from $I(S)$ we should recognize if this net implements the function $f_1 = f$. The *problem of diagnosis*: for a given net from $I(S)$ we should recognize a function f_i, $i \in \{1,\dots,m\}$, which this net implements. To resolve these problems we can give arbitrary tuples of values from the set $\{0,1\}^n$ to the inputs of net and observe the values at the output of the net.

We associate two decision tables with the considered problems. Both tables have m rows corresponding to functions $f_1,\dots,f_m$ and 2^n columns corresponding to tuples from $\{0,1\}^n$. At the intersection with row f_i and column $\alpha \in \{0,1\}^n$ the value $f_i(\alpha)$ stays. If we label the row f_1 with the decision 1 and the rows $f_2,\dots,f_m$ with the decision 0, we obtain *control* decision table $T_C(S,I)$. If we label the row f_i, $i = 1,\dots,m$, with the decision i, we obtain *diagnostic* decision table $T_D(S,I)$.

We will study tests and decision trees for tables $T_C = T_C(S,I)$ and $T_D = T_D(S,I)$. Tests for T_C are called *control* tests. One can show that the minimum cardinality $R(T_C)$ of control test for T_C is equal to the minimum depth $h(T_C)$ of a decision tree for T_C, i.e., $R(T_C) = h(T_C)$. Tests for T_D are called *diagnostic* tests.

1.5.2 Contact Networks and Combinatorial Circuits

We describe here the notions of contact network and combinatorial circuit, and list some types of faults.

1.5.2.1 Contact Networks

A *contact network* is a finite undirected graph S such that edges of this graph are labeled with literals: variables $x_1,\dots,x_n$ or negations of these variables $\neg x_1,\dots,\neg x_n$. Two nodes of this graph s and t are fixed as *input* and *output* nodes respectively. The considered contact network S implements a Boolean function $f(x_1,\dots,x_n)$. This function is equal to 1 on a tuple $\alpha \in \{0,1\}^n$ of values of variables $x_1,\dots,x_n$ if and only if there exists a path from s to t such that, for every edge of the path, the literal attached to the edge is equal to 1 for the tuple α.

We consider *constant faults* for contact networks when instead of a literal attached to an edge we have a constant from the set $\{0,1\}$. We distinguish *single* faults when in the network we can have at most one edge with fault, and *complete* faults when faults can be in arbitrary number of edges.

1.5.2.2 Combinatorial Circuits

A *basis* is an arbitrary nonempty finite set of Boolean functions. Let B be a basis.

A *combinatorial circuit in the basis B* (a *circuit in the basis B*) is a labeled finite directed acyclic graph with multiple edges which has nodes of the three types: inputs, gates and outputs.

Nodes of the *input* type have no entering edges, each input is labeled with a variable, and distinct inputs are labeled with distinct variables. Every circuit has at least one input.

Each node of the *gate* type is labeled with a function from the set B. Let v be a gate and let a function g depending on t variables be attached to it. If $t = 0$ (this is the case when g is one of the constants 0 or 1) then the node v has no entering edges. If $t > 0$ then the node v has exactly t entering edges which are labeled with numbers $1,\dots,t$ respectively. Every circuit has at least one gate.

Each node of the *output* type has exactly one entering edge which issues from a gate. Let v be an output. Nothing is attached to v, and v has no issuing edges. We will consider only circuits which have exactly one output.

Let S be a circuit in the basis B which has n inputs labeled with variables $x_1,\dots,$ x_n. Let us correspond to each node v in the circuit S a Boolean function f_v depending on variables $x_1,\dots,x_n$. If v is an input of S labeled with the variable x_i then $f_v = x_i$. If v is a gate labeled with a constant $c \in \{0,1\}$ then $f_v = c$. Let v be a gate labeled with a function g depending on $t > 0$ variables. For $i = 1,\dots,t$, let the edge d_i, labeled with the number i, issue from a node v_i and enter the node v. Then $f_v = g(f_{v_1},\dots,f_{v_t})$. If v is an output of the circuit S and an edge, issuing from a node u, enters the node v, then $f_v = f_u$. The function corresponding to the output of the circuit S will be denoted by f. We will say that the circuit S *implements* the function f.

For combinatorial circuit S in the basis B, we can consider a new basis P for faults and can build in (under some constrains) gates implementing functions from P into the circuit S. We will assume later that each function from P depends essentially on each its variable. The most known bases for faults are $\{0\}$, $\{1\}$, $\{0,1\}$, and $\{\neg x\}$. We can consider also constant faults 0 and 1 on inputs of gates in S, or negation faults $\neg$ on outputs of gates in S, etc. As in the case of contact networks, we distinguish *single* and *complete* faults.

We now consider an example from [42].

Example 1.86. Let S be a combinatorial circuit with one gate represented in Fig. 1.7(a). We will admit constant 0 and 1 faults on gate inputs. The circuit S (possible, with faults) implements a function from the set $\{x \wedge y, x, y, 0, 1\}$. In Fig. 1.7(b) one can see corresponding diagnostic decision table and in Fig. 1.7(c) one can see a decision tree which solves the diagnosis problem for S. Note that in the table and in the tree we use functions as labels instead of its numbers.

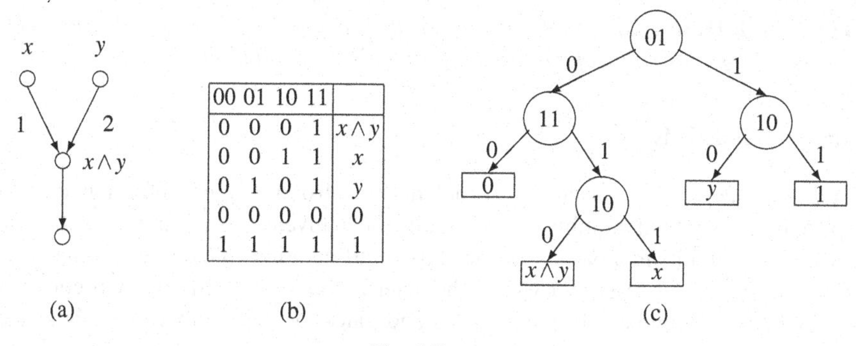

Fig. 1.7.

1.5.3 Faults on Inputs

We begin our consideration of results from faults on inputs of nets. Let we have a net S with n inputs $x_1,\ldots,x_n$ which implements a Boolean function $f(x_1,\ldots,x_n)$, and a source of faults I which affects only on inputs of S. This source can put, for example, constants from $\{0,1\}$ instead of some inputs (variables). In this case, problems of control and diagnosis do not depend on the structure of the net S, but only on the function f and faults I. So we will say about tests for a Boolean function.

First, we consider diagnostic tests for constant (both 0 and 1) faults on inputs.

Theorem 1.87. *(Noskov [50]) For any Boolean function with n variables, there exists a diagnostic test for complete constant faults which cardinality is at most* $4(n+1)^3 2^{0.773n}$. *For any natural n, there exists a Boolean function with n variables for which the minimum cardinality of diagnostic test for complete constant faults is at least* $2^{0.5n-1}/\sqrt{n}$.

Theorem 1.88. *(Noskov [52]) For any Boolean function with n variables, there exists a diagnostic test for single constant faults which cardinality is at most* $2n$. *For any natural n, there exists a Boolean function with n variables for which the minimum cardinality of diagnostic test for single constant faults is at least* $2n$.

Theorem 1.89. *(Noskov [52]) For almost all Boolean functions with n variables, the minimum cardinality of diagnostic test for single constant faults is greater than* $\log_2 n$ *and less than* $17\log_2 n$.

We consider now control tests for constant (both 0 and 1) faults on inputs.

For $n \geq 136$, we define the function $L(n)$ as follows:

$$L(n) = \begin{cases} 2n-2t-1, & \text{if } n = 2^t+t+1\,, \\ 2n-2t-2, & \text{if } 2^t+t+1 < n \leq 2^{t+1}+t+1\,. \end{cases}$$

Theorem 1.90. *(Noskov [51]) For any Boolean function with* $n \geq 136$ *variables, there exists a control test for complete constant faults which cardinality is at most* $L(n)$. *For any* $n \geq 136$, *there exists a Boolean function with n variables for which the minimum cardinality of control test for complete constant faults is at least* $L(n)$.

Theorem 1.91. *(Noskov [51]) For almost all Boolean functions with n variables, the minimum cardinality of control test for single constant faults is at most* 3.

1.5.4 Net as It Is

In this subsection, we consider the problem of control or diagnosis for a given (not specially constructed) combinatorial circuit from a given class of circuits (either the class of all circuits in a basis B or the class of all iteration-free circuits in the basis B). As a parameter of circuit which characterizes its complexity we will consider either the total number of circuit inputs and gates, or the number of gates in the circuit.

The most general results in this direction were obtained by Shevtchenko [61]. Let B be an arbitrary basis for circuits and P be an arbitrary basis for faults which consist of embedding of gates implementing functions from P into a circuit S in the basis B. We denote by $h^d_P(S)$ the minimum depth of a decision tree which solves the problem of diagnosis of S relative to complete faults from P. For every natural $n \geq 2$, we denote by $h^d_{B,P}(n)$ the maximum value of $h^d_P(S)$ where maximum is given among all circuits in the basis B with total number of circuit inputs and gates at most n.

A finite set F of Boolean functions will be called *primitive* if at least one of the following conditions holds:

a) every function from F is either a disjunction $x_1 \vee \ldots \vee x_n$ or a constant;
b) every function from F is either a conjunction $x_1 \wedge \ldots \wedge x_n$ or a constant;
c) every function from F is either a linear function $x_1 + \ldots + x_n + c \pmod 2$, $c \in \{0,1\}$, or a constant.

Theorem 1.92. *(Shevtchenko [61]) For every circuit basis B and every basis of faults P the following statements hold:*

a) if $B \cup P$ is a primitive set then, for any natural $n \geq 2$, either

$$0 \leq h^d_{B,P}(n) \leq 2$$

or

$$\lceil (n+1)/2 \rceil \leq h^d_{B,P}(n) \leq n\,;$$

b) if $B \cup P$ is a non-primitive set then, for any natural $n \geq 2$,

$$2^{\lfloor (n+3)/7 \rfloor} \leq h^d_{B,P}(n) \leq 2^{n-1}\,.$$

Analogous study was done by Shevtchenko for the problem of control. We consider here one of his results for negation type faults. Let B be a circuit basis and S be a circuit in the basis B. We denote by $h^c_\neg(S)$ the minimum depth of a decision tree which solves the problem of control of S relative to complete negation faults on inputs and outputs of gates. Note that $h^c_\neg(S)$ is equal to $R^c_\neg(S)$ which is the minimum cardinality of a control test for S relative to complete negation faults on inputs and outputs of gates. For every natural n, we denote by $h^c_{B,\neg}(n)$ the maximum value of $h^c_\neg(S)$ where maximum is given among all circuits in the basis B with at most n gates.

Theorem 1.93. *(Shevtchenko [60]) For every basis B the following statements hold:*

a) if B contains only linear functions then, for any natural n,

$$h^c_{B,\neg}(n) = 1 ;$$

b) if B contains at least one function which is not linear then, for any natural n,

$$2^{\lfloor (n-1)/2 \rfloor} - 1 \le h^c_{B,\neg}(n) \le 2^{(r-1)n+1}$$

where r is the maximum number of variables among functions from B.

The next result is connected with constant faults (both 0 and 1) on inputs of gates of a circuit S in a basis B. We denote by $h^d_{\{0,1\}}(S)$ the minimum depth of a decision tree which solves the problem of diagnosis of S relative to complete constant faults on inputs of gates from S. For every natural n, we denote by $h^d_{B,\{0,1\}}(n)$ the maximum value of $h^d_{\{0,1\}}(S)$ where maximum is given among all circuits in the basis B with number of gates at most n.

Theorem 1.94. *(Moshkov [41]) For every basis B, the following statements hold:*
a) if B is a primitive set then

$$h^d_{B,\{0,1\}}(n) = O(n) ;$$

b) if B is a non-primitive set then

$$\log_2 h^d_{B,\{0,1\}}(n) = \Omega(n^{1/2}) .$$

We consider now similar results but obtained for iteration-free circuits. A circuit in the basis B is called *iteration-free* if each node (input or gate) of it has at most one issuing edge. For every natural n, we denote by $h^{d,i-f}_{B,\{0,1\}}(n)$ the maximum value of $h^d_{\{0,1\}}(S)$ where maximum is given among all iteration-free circuits in the basis B with number of gates at most n.

From the results obtained by Goldman and Chipulis in [22], and Karavai in [27] the bound $h^{d,i-f}_{B,\{0,1\}}(n) = O(n)$ can be derived immediately for arbitrary basis B with the following property: each function from B is implemented by some iteration-free circuit in the basis $\{x \wedge y, x \vee y, \neg x\}$. We consider now a generalization of this result.

We will say that a Boolean function $f(x_1,\dots,x_n)$ is *quasimonotone* if there exist numbers $\sigma_1,\dots, \sigma_n \in \{0,1\}$ and a monotone Boolean function $g(x_1,\dots,x_n)$ such that $f(x_1,\dots,x_n) = g(x_1^{\sigma_1},\dots,x_n^{\sigma_n})$ where $x^\sigma = x$ if $\sigma = 1$, and $x^\sigma = \neg x$ if $\sigma = 0$.

A finite set of Boolean functions F will be called *quasiprimitive* if at least one of the following conditions is true:

a) all functions from F are linear functions or constants;
b) all functions from F are quasimonotone functions.

The class of quasiprimitive bases (which are quasiprimitive sets) is rather large: for any basis B_1 there exists a quasiprimitive basis B_2 such that the set of Boolean functions implemented by circuits in the basis B_1 coincides with the set of Boolean functions implemented by circuits in the basis B_2.

Theorem 1.95. *(Moshkov [41]) For every basis B, the following statements hold:*

a) if B is a quasiprimitive set then

$$h^{d,i-f}_{B,\{0,1\}}(n) = O(n) ;$$

b) if B is a non-quasiprimitive set then

$$\log_2 h^{d,i-f}_{B,\{0,1\}}(n) = \Omega(n) .$$

Results about efficient diagnosis of iteration-free circuits relative to so-called retaining faults were obtained by Moshkova in [46, 47].

For a given (arbitrary) circuit S in a basis B we can have problems even in construction of arbitrary (not optimal) decision tree for diagnosis of constant faults on inputs of gates.

If we fix some inputs of gates and constant faults on these inputs we will obtain a *tuple* of constant faults on inputs of gates of the circuit S.

We define now a *problem* $\mathrm{Con}(B)$: for a given circuit S in the basis B and a given set W of tuples of constant faults on inputs of gates of the circuit S it is required to construct a decision tree which solves the diagnosis problem for the circuit S relative to the faults from W.

Note that there exists a decision tree which solves the diagnosis problem for the circuit S relative to the faults from W and the number of nodes in which is at most $2|W|-1$.

Theorem 1.96. *(Moshkov [41]) Let B be a basis such that $B \not\subseteq \{0,1\}$. Then the following statements hold:*

a) if B is a primitive set then there exists an algorithm which solves the problem $\mathrm{Con}(B)$ *with polynomial time complexity;*

b) if B is a non-primitive set then the problem $\mathrm{Con}(B)$ *is NP-hard.*

1.5.5 Specially Constructed Nets

The most part of results connected with fault control and diagnosis is about specially constructed contact networks and combinatorial circuits.

In this subsection, very often we will study so-called Shannon functions. Let us consider, for example, a Shannon function $R^d_{0,1}(n)$ corresponding to the cardinality of tests for the problem of diagnosis of contact networks relative to complete constant faults. For a contact network S, we denote by $R(S)$ the minimum cardinality of a test for the problem of diagnosis of S relative to complete constant faults. For a given Boolean function f, we denote by $R(f)$ the minimum value of $R(S)$ where the minimum is given among all contact networks S implementing f. For every natural n, the value of Shannon function $R^d_{0,1}(n)$ is the maximum value of $R(f)$ where the maximum is given among all Boolean functions f with n variables.

Instead of contact networks we can consider combinatorial circuits. In this case, we will fix a circuit basis B.

1.5.5.1 Specially Constructed Contact Networks

We will begin from a result relative to the problem of diagnosis.

Theorem 1.97. *(Madatyan [35]) Let $R^d_{0,1}(n)$ be the Shannon function corresponding to the cardinality of tests for the problem of diagnosis of contact networks relative to complete constant faults. Then*

$$R^d_{0,1}(n) = 2^n \, .$$

The rest of results is connected with the problem of control. Later, for two functions $a(n)$ and $b(n)$, we will use the notation $a(n) \lesssim b(n)$ to describe the fact that

$$\limsup_{n \to \infty} \frac{a(n)}{b(n)} \leq 1$$

or which is the same

$$\overline{\lim}_{n \to \infty} \frac{a(n)}{b(n)} \leq 1 \, .$$

Theorem 1.98. *(Chegis and Yablonskii [10]) Let $\sigma \in \{0,1\}$ and $R^{c,s}_{\sigma}(n)$ be the Shannon function corresponding to the cardinality of tests for the problem of control of contact networks relative to single σ faults. Then*

$$R^{c,s}_{\sigma}(n) \lesssim \frac{2^n}{n} \, .$$

This bound was improved by Madatyan:

Theorem 1.99. *(Madatyan [36]) Let $\sigma \in \{0,1\}$ and $R^{c,s}_{\sigma}(n)$ be the Shannon function corresponding to the cardinality of tests for the problem of control of contact networks relative to single σ faults. Then*

$$R^{c,s}_{\sigma}(n) \lesssim \frac{2^n}{n\sqrt{n}} \, .$$

Theorem 1.100. *(Red'kin [55]) Let $R^c_{0,1}(n)$ be the Shannon function corresponding to the cardinality of tests for the problem of control of contact networks relative to complete constant faults. Then, for $n \geq 4$,*

$$R^c_{0,1}(n) \leq \frac{15}{6} 2^n .$$

Theorem 1.101. *(Red'kin [56]) Let $R_0^c(n)$ be the Shannon function corresponding to the cardinality of tests for the problem of control of contact networks relative to complete 0 faults. Then*

$$R_0^c(n) \le 2^{\lfloor n/2 \rfloor} + 2^{\lceil n/2 \rceil} .$$

Theorem 1.102. *(Red'kin [56]) Let $R_1^c(n)$ be the Shannon function corresponding to the cardinality of tests for the problem of control of contact networks relative to complete 1 faults. Then*

$$R_1^c(n) \lesssim 2^{\frac{n}{1+0.5\log_2 n}+2.5} .$$

1.5.5.2 Specially Constructed Combinatorial Circuits

A basis B is called *complete* if any Boolean function can be implemented by a circuit in the basis B.

Theorem 1.103. *(Red'kin [57]) Let B be a complete basis and $R_{B,\{0,1\}}^d(n)$ be the Shannon function corresponding to the cardinality of tests for the problem of diagnosis of combinatorial circuits in the basis B relative to complete constant faults on the outputs of gates. Then*

$$R_{B,\{0,1\}}^d(n) \le 2\left(2^{\lfloor n/2 \rfloor} + 2^{\lceil n/2 \rceil} + n\right) .$$

Theorem 1.104. *(Red'kin [58]) Let B be a complete basis and $R_{B,\neg}^{c,s}(n)$ be the Shannon function corresponding to the cardinality of tests for the problem of control of combinatorial circuits in the basis B relative to single negation faults on the outputs of gates. Then*

$$R_{B,\neg}^{c,s}(n) \le 3 .$$

Theorem 1.105. *(Borodina [7, 8, 9]) Let $B = \{x \wedge y, x \vee y, \neg x\}$, $\sigma \in \{0,1\}$ and $R_{B,\sigma}^c(n)$ be the Shannon function corresponding to the cardinality of tests for the problem of control of combinatorial circuits in the basis B relative to complete σ faults on the outputs of gates. Then*

$$R_{B,\sigma}^c(n) = 2 .$$

1.6 Discrete Optimization and Recognition of Words

In this section, we consider two less known areas of applications of TT connected with discrete optimization and mathematical linguistics. This section is based mainly on monographs [42, 45]

1.6.1 Discrete Optimization

This subsection is devoted to the consideration of three classes of discrete optimization problems over quasilinear information systems. For each class, examples and corollaries of Theorem 1.82 are given. More detailed discussion of considered results can be found in [39, 42].

1.6.1.1 Traveling Salesman Problem with Four Cities

We begin with example of traveling salesman problem for four cities [45] which illustrates TT approach to the study of discrete optimization problems.

Let we have complete unordered graph with four nodes in which each edge is marked by a real number – the length of this edge (see Fig. 1.8).

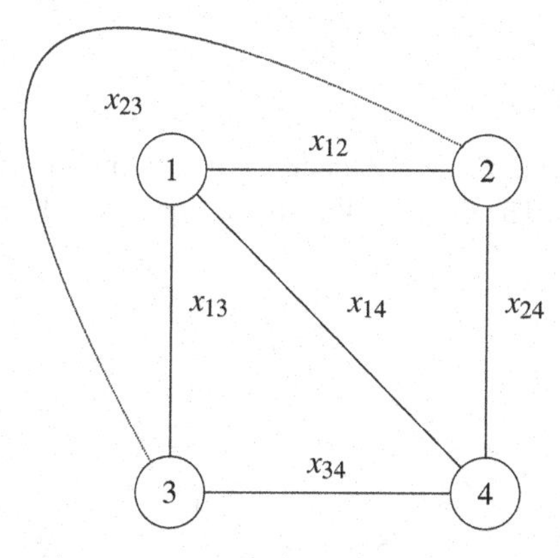

Fig. 1.8.

A Hamiltonian circuit is a closed path which passes through each node exactly one time. There are three Hamiltonian circuits:

H_1: 12341 or, which is the same, 14321,
H_2: 12431 or 13421,
H_3: 13241 or 14231.

We should find the number d of a Hamiltonian circuit which has the minimum length. For $i = 1,2,3$, we denote by L_i the length of H_i. Then

$$L_1 = x_{12} + x_{23} + x_{34} + x_{14} = \overbrace{(x_{12} + x_{34})}^{\alpha} + \overbrace{(x_{23} + x_{14})}^{\beta}\,,$$
$$L_2 = x_{12} + x_{24} + x_{34} + x_{13} = \overbrace{(x_{12} + x_{34})}^{\alpha} + \overbrace{(x_{24} + x_{13})}^{\gamma}\,,$$
$$L_3 = x_{13} + x_{23} + x_{24} + x_{14} = \overbrace{(x_{24} + x_{13})}^{\gamma} + \overbrace{(x_{23} + x_{14})}^{\beta}\,.$$

For simplicity, we will assume that L_1, L_2 and L_3 are pairwise different numbers. So, as universe we will consider the set of points of the space $\mathbb{R}^6$ which do not lie on hyperplanes defined by equations $L_1 = L_2$, $L_1 = L_3$, $L_2 = L_3$. In the capacity of attributes we will use three functions $f_1 = \text{sign}(L_1 - L_2)$, $f_2 = \text{sign}(L_1 - L_3)$, and $f_3 = \text{sign}(L_2 - L_3)$.

Values L_1, L_2 and L_3 are linearly ordered. Any order is possible since values of α, β and γ can be chosen independently – see Fig. 1.9 which contains also corresponding decision table T.

If $\alpha < \beta < \gamma$ then $L_1 < L_2 < L_3$
If $\alpha < \gamma < \beta$ then $L_2 < L_1 < L_3$
If $\beta < \alpha < \gamma$ then $L_1 < L_3 < L_2$
If $\beta < \gamma < \alpha$ then $L_3 < L_1 < L_2$
If $\gamma < \alpha < \beta$ then $L_2 < L_3 < L_1$
If $\gamma < \beta < \alpha$ then $L_3 < L_2 < L_1$

f_1	f_2	f_3	d
-1	-1	-1	1
$+1$	-1	-1	2
-1	-1	$+1$	1
-1	$+1$	$+1$	3
$+1$	$+1$	-1	2
$+1$	$+1$	$+1$	3

$= T$

Fig. 1.9.

It is easy to show that $h(T) \geq 2$. A decision tree, represented in Fig. 1.10 solves the considered problem. The depth of this tree is equal to 2. Hence $h(T) = 2$.

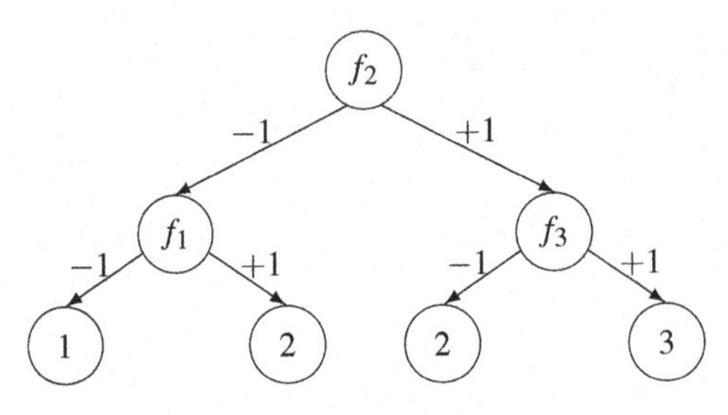

Fig. 1.10.

1.6.1.2 Some Definitions

Let $U = U(A, K, \varphi_1, \ldots, \varphi_m)$ be a quasilinear information system (see Sect. 1.3 for definitions). A pair (A, ϕ) where ϕ is a function from A to a finite subset of the set $\omega = \{0, 1, 2, \ldots\}$ will be called a *problem over* A. The problem (A, ϕ) may be interpreted as a problem of searching for the value $\phi(a)$ for a given $a \in A$. Let $k \in \omega$, $k \geq 1$, and $t \in \mathbb{R}$, $t \geq 0$. The problem (A, ϕ) will be called (m,k,t)*-problem over* U if there exists a problem z over U such that $\phi(a) = z(a)$ for each $a \in A$, $\dim z \leq k$ and $r(z) \leq t$. Let $\phi(a) = z(a)$ for each $a \in A$ and $z = (\nu, f_1, \ldots, f_p)$. Then the set $\{f_1, \ldots, f_p\}$ will be called a *separating set* (*with attributes from* $F(A, K, \varphi_1, \ldots, \varphi_m)$) for the problem (A, ϕ). We will say that a decision tree Γ over U *solves* the problem (A, ϕ) if the decision tree Γ solves the problem z.

We denote

$$L(A, K, \varphi_1, \ldots, \varphi_m) = \left\{ \sum_{i=1}^{m} d_i \varphi_i(x) + d_{m+1} : d_1, \ldots, d_{m+1} \in K \right\} .$$

Let $g \in L(A, K, \varphi_1, \ldots, \varphi_m)$ and $g = \sum_{i=1}^{m} d_i \varphi_i(x) + d_{m+1}$. We define the parameter $r(g)$ of the function g as follows. If $(d_1, \ldots, d_{m+1}) = (0, \ldots, 0)$ then $r(g) = 0$. Otherwise

$$r(g) = \max\{0, \max\{\log_2 |d_i| : i \in \{1, \ldots, m+1\}, d_i \neq 0\}\} .$$

In what follows we will assume that elements of the set $\{-1,1\}^n$ and of the set $\{0,1\}^n$ are enumerated by numbers from 1 to 2^n.

1.6.1.3 Problems of Unconditional Optimization

Let $k \in \omega \setminus \{0\}$, $t \in \mathbb{R}, t \geq 0$, $g_1,\ldots,g_k \in L(A,K,\varphi_1,\ldots,\varphi_m)$, and $r(g_j) \leq t$ for $j = 1,\ldots,k$.

Problem 1.106. (Unconditional optimization of values of functions $g_1,\ldots,g_k$ on an element of the set A.) For a given $a \in A$, it is required to find the minimum number $i \in \{1,\ldots,k\}$ such that $g_i(a) = \min\{g_j(a) : 1 \leq j \leq k\}$.

One can show that the set $\{\text{sign}(g_i(x) - g_j(x)) : i,j \in \{1,\ldots,k\}, i \neq j\}$ is a separating set for this problem, and the considered problem is $(m,k^2,t+1)$-problem over the information system $U(A,K,\varphi_1,\ldots,\varphi_m)$.

Example 1.107. (n-City traveling salesman problem.) Let $n \in \omega$, $n \geq 4$, and let G_n be the complete undirected graph with n nodes. Assume that edges in G_n are enumerated by numbers from 1 to $n(n-1)/2$, and Hamiltonian circuits in G_n are enumerated by numbers from 1 to $(n-1)!/2$. Let a number $a_i \in \mathbb{R}$ be attached to the i-th edge, $i = 1,\ldots,n(n-1)/2$. We will interpret the number a_i as the length of the i-th edge. It is required to find the minimum number of a Hamiltonian circuit in G_n which has the minimum length. For each $j \in \{1,\ldots,(n-1)!/2\}$, we will associate with the j-th Hamiltonian circuit the function $g_j(\bar{x}) = \sum_{i=1}^{n(n-1)/2} \delta_{ji} x_i$ where $\delta_{ji} = 1$ if the i-th edge is contained in the j-th Hamiltonian circuit, and $\delta_{ji} = 0$ otherwise. Obviously, the considered problem is the problem of unconditional optimization of values of functions $g_1,\ldots,g_{(n-1)!/2}$ on an element of the set $\mathbb{R}^{n(n-1)/2}$. Therefore the set $\{\text{sign}(g_i(\bar{x}) - g_j(\bar{x})) : i,j \in \{1,\ldots,(n-1)!/2\}, i \neq j\}$ is a separating set for the n-city traveling salesman problem, and this problem is $(n(n-1)/2, ((n-1)!/2)^2, 0)$-problem over the information system $U = U(\mathbb{R}^{n(n-1)/2}, \mathbb{Z}, x_1,\ldots,x_{n(n-1)/2})$. From Theorem 1.82 it follows that there exists a decision tree Γ over U which solves the n-city traveling salesman problem and for which $h(\Gamma) \leq n^7/2$ and $r(\Gamma) \leq n^4 \log_2 n$.

1.6.1.4 Problems of Unconditional Optimization of Absolute Values

Let $k \in \omega \setminus \{0\}$, $t \in \mathbb{R}, t \geq 0$, $g_1,\ldots,g_k \in L(A,K,\varphi_1,\ldots,\varphi_m)$, and $r(g_j) \leq t$ for $j = 1,\ldots,k$.

Problem 1.108. (Unconditional optimization of absolute values of functions $g_1,\ldots,g_k$ on an element of the set A.) For a given $a \in A$, it is required to find the minimum number $i \in \{1,\ldots,k\}$ such that $|g_i(a)| = \min\{|g_j(a)| : 1 \leq j \leq k\}$.

Evidently, $|g_i(a)| < |g_j(a)|$ if and only if $(g_i(a) + g_j(a))(g_i(a) - g_j(a)) < 0$, and $|g_i(a)| = |g_j(a)|$ if and only if $(g_i(a) + g_j(a))(g_i(a) - g_j(a)) = 0$. Using these relations one can show that the set $\{\text{sign}(g_i(x) + g_j(x)), \text{sign}(g_i(x) - g_j(x)) : i,j \in \{1,\ldots,k\}, i \neq j\}$ is a separating set for the considered problem, and this problem is $(m,2k^2,t+1)$-problem over the information system $U(A,K,\varphi_1,\ldots,\varphi_m)$.

Example 1.109. (n-Stone problem.) Let $n \in \omega \setminus \{0\}$. For a tuple $(a_1,\dots,a_n) \in \mathrm{I\!R}^n$, it is required to find the minimum number of a tuple $(\delta_1,\dots,\delta_n) \in \{-1,1\}^n$ which minimizes the value of $|\sum_{i=1}^{n} \delta_i a_i|$. Obviously, this problem is the problem of unconditional optimization of absolute values of functions from the set $\{\sum_{i=1}^{n} \delta_i x_i : (\delta_1,\dots,\delta_n) \in \{-1,1\}^n\}$ on an element of the set $\mathrm{I\!R}^n$. Therefore the set

$$\{\mathrm{sign}(\sum_{i=1}^{n} \delta_i x_i) : (\delta_1,\dots,\delta_n) \in \{-2,0,2\}^n\}$$

and, hence, the set

$$\{\mathrm{sign}(\sum_{i=1}^{n} \delta_i x_i) : (\delta_1,\dots,\delta_n) \in \{-1,0,1\}^n\}$$

are separating sets for the considered problem, and this problem is $(n,3^n,0)$-problem over the information system $U = U(\mathrm{I\!R}^n, \mathbb{Z}, x_1,\dots,x_n)$. From Theorem 1.82 it follows that there exists a decision tree Γ over U which solves the n-stone problem and for which $h(\Gamma) \leq 4(n+2)^4/\log_2(n+2)$ and $r(\Gamma) \leq 2(n+1)^2\log_2(2n+2)$.

1.6.1.5 Problems of Conditional Optimization

Let $k,p \in \omega \setminus \{0\}$, $t \in \mathrm{I\!R}, t \geq 0$, $D \subseteq \mathrm{I\!R}$, $D \neq \emptyset$ and $g_1,\dots,g_k$ be functions from $L(A,K,\varphi_1,\dots,\varphi_m)$ such that $r(g_j) \leq t$ for $j = 1,\dots,k$.

Problem 1.110. (Conditional optimization of values of functions $g_1,\dots,g_k$ on an element of the set A with p restrictions from $A \times D$.) For a given tuple $(a_0,a_1,\dots,a_p,d_1,\dots,d_p) \in A^{p+1} \times D^p$, it is required to find the minimum number $i \in \{1,\dots,k\}$ such that $g_i(a_1) \leq d_1,\dots,g_i(a_p) \leq d_p$ and

$$g_i(a_0) = \max\{g_j(a_0) : g_j(a_1) \leq d_1,\dots,g_j(a_p) \leq d_p, j \in \{1,\dots,k\}\}$$

or to show that such i does not exist. (In the last case let $k+1$ be the solution of the problem.)

The variables with values from A will be denoted by $x_0,x_1,\dots,x_p$ and the variables with values from D will be denoted by $y_1,\dots,y_p$. One can show that the set $\{\mathrm{sign}(g_i(x_0) - g_j(x_0)) : 1 \leq i,j \leq k\} \cup (\bigcup_{j=1}^{p}\{\mathrm{sign}(g_i(x_j) - y_j) : 1 \leq i \leq k\})$ is a separating set for the considered problem, and this problem is $(p+m(p+1), pk+k^2, t+1)$-problem over the information system

$$U(A^{p+1} \times D^p, K, \varphi_1(x_0),\dots,\varphi_m(x_0),\dots,\varphi_1(x_p),\dots,\varphi_m(x_p),y_1,\dots,y_p).$$

Example 1.111. (Problem on 0-1-knapsack with n objects.) Let $n \in \omega \setminus \{0\}$. For a given tuple $(a_1,\dots,a_{2n+1}) \in \mathbb{Z}^{2n+1}$, it is required to find the minimum number of a tuple $(\delta_1,\dots,\delta_n) \in \{0,1\}^n$ which maximizes the value $\sum_{i=1}^{n} \delta_i a_i$ under the condition $\sum_{i=1}^{n} \delta_i a_{n+i} \leq a_{2n+1}$. This is the problem of conditional optimization of values of functions from the set $\{\sum_{i=1}^{n} \delta_i x_i : (\delta_1,\ldots,\delta_n) \in \{0,1\}^n\}$ on an element of the set $\mathbb{Z}^n$ with one restriction from $\mathbb{Z}^n \times \mathbb{Z}$. The set $\{\mathrm{sign}(\sum_{i=1}^{n} \delta_i x_i) : (\delta_1,\ldots,\delta_n) \in \{-1,0,1\}^n\} \cup \{\mathrm{sign}(\sum_{i=1}^{n} \delta_i x_{n+i} - x_{2n+1}) : (\delta_1,\ldots,\delta_n) \in \{0,1\}^n\}$ is a separating set for the considered problem, and this problem is $(2n+1, 3^n+2^n, 0)$-problem over the information system $U = U(\mathbb{Z}^{2n+1}, \mathbb{Z}, x_1, \ldots, x_{2n+1})$. From Theorem 1.82 it follows that there exists a decision tree Γ over U which solves the problem on 0-1-knapsack with n objects and for which $h(\Gamma) \leq 2(2n+3)^4/\log_2(2n+3)$ and $r(\Gamma) \leq 2(2n+2)^2 \log_2(4n+4)$.

1.6.2 Regular Language Word Recognition

In this subsection, we consider the problem of recognition of words of fixed length in a regular language. The word under consideration can be interpreted as a description of certain screen image in the following way: the i-th letter of the word encodes the color of the i-th screen cell. In this case, a decision tree (or a decision rule system) which recognizes some words may be interpreted as an algorithm for the recognition of images which are defined by these words. The considered here results (mainly with proofs) can be found in [40, 42].

1.6.2.1 Problem of Recognition of Words

Let $k \in \omega$, $k \geq 2$ and $E_k = \{0,1,\ldots,k-1\}$. By $(E_k)^*$ we denote the set of all finite words over the alphabet E_k, including the empty word λ. Let $\mathscr{L}$ be a regular language over the alphabet E_k. For $n \in \omega \setminus \{0\}$, we denote by $\mathscr{L}(n)$ the set of all words from $\mathscr{L}$ for which the length is equal to n. Let us assume that $\mathscr{L}(n) \neq \emptyset$. For $i \in \{1,\ldots,n\}$, we define a function $l_i : \mathscr{L}(n) \to E_k$ as follows: $l_i(\delta_1 \ldots \delta_n) = \delta_i$ for each $\delta_1 \ldots \delta_n \in \mathscr{L}(n)$. Let us consider an information system $U(\mathscr{L},n) = (\mathscr{L}(n), E_k, \{l_1,\ldots,l_n\})$ and a problem $z_{\mathscr{L},n} = (\nu, l_1,\ldots,l_n)$ over $U(\mathscr{L},n)$ such that $\nu(\bar{\delta}_1) \neq \nu(\bar{\delta}_2)$ for every $\bar{\delta}_1, \bar{\delta}_2 \in E_k^n$, $\bar{\delta}_1 \neq \bar{\delta}_2$. The problem $z_{\mathscr{L},n}$ will be called the *problem of recognition of words from* $\mathscr{L}(n)$.

We denote by $h_{\mathscr{L}}(n)$ the minimum depth of a decision tree over $U(\mathscr{L},n)$ which solves the problem of recognition of words from $\mathscr{L}(n)$. If $\mathscr{L}(n) = \emptyset$ then $h_{\mathscr{L}}(n) = 0$. We denote by $L_{\mathscr{L}}(n)$ the minimum depth of a decision rule system over $U(\mathscr{L},n)$ which is complete for the problem of recognition of words from $\mathscr{L}(n)$. If $\mathscr{L}(n) = \emptyset$ then $L_{\mathscr{L}}(n) = 0$.

Here, we consider the behavior of two functions $H_{\mathscr{L}} : \omega \setminus \{0\} \to \omega$ and $P_{\mathscr{L}} : \omega \setminus \{0\} \to \omega$ which are defined as follows. Let $n \in \omega \setminus \{0\}$. Then

$$H_{\mathscr{L}}(n) = \max\{h_{\mathscr{L}}(m) : m \in \omega \setminus \{0\}, m \leq n\},$$
$$P_{\mathscr{L}}(n) = \max\{L_{\mathscr{L}}(m) : m \in \omega \setminus \{0\}, m \leq n\}.$$

Example 1.112. Let $\mathscr{L}$ be the regular language which is generated by the source represented in Fig. 1.11 (the node q_0 is labeled with $+$, and the unique node from Q is labeled with $*$). Let us consider the problem $z_{\mathscr{L},4} = (\nu, l_1, l_2, l_3, l_4)$ of recognition of words from $\mathscr{L}(4) = \{1110, 1100, 1000, 0000\}$. Let $\nu(1,1,1,0) = 1$,

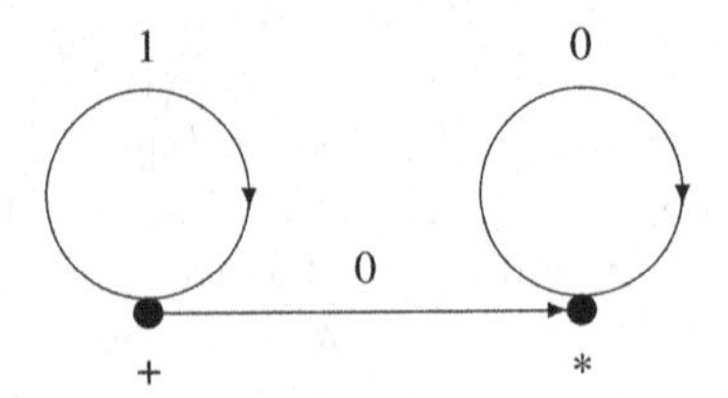

Fig. 1.11.

$\nu(1,1,0,0)=2$, $\nu(1,0,0,0)=3$ and $\nu(0,0,0,0)=4$. The decision table $T(z_{\mathscr{L},4})$ is represented in Fig. 1.12(a). The decision tree in Fig. 1.12(b) solves the problem of recognition of words from $\mathscr{L}(4)$. Note that instead of numbers of words the terminal nodes in this tree are labeled with words. The depth of the considered decision tree is equal to 2. Using Theorem 1.7, we obtain $h_{\mathscr{L}}(4)=2$.

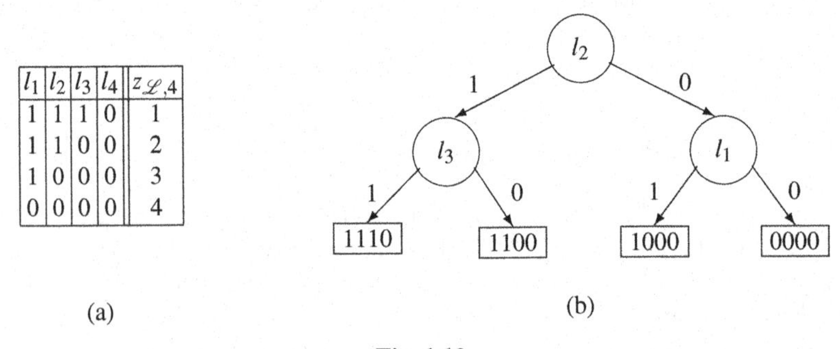

l_1	l_2	l_3	l_4	$z_{\mathscr{L},4}$
1	1	1	0	1
1	1	0	0	2
1	0	0	0	3
0	0	0	0	4

Fig. 1.12.

The decision rule system

$$\{l_3=1\rightarrow 1, l_2=1\wedge l_3=0\rightarrow 2, l_1=1\wedge l_2=0\rightarrow 3, l_1=0\rightarrow 4\}$$

is complete for the problem of recognition of words from $\mathscr{L}(4)$. The depth of this system is equal to 2. One can show that $M(T(z_{\mathscr{L},4}),(1,1,0,0))=2$. Using Theorem 1.16, we obtain $L_{\mathscr{L}}(4)=2$.

1.6.2.2 A-Sources

An *A-source over the alphabet* E_k is a triple $I=(G,q_0,Q)$ where G is a directed graph, possibly with multiple edges and loops, in which each edge is labeled with a number from E_k and any edges starting in a node are labeled with pairwise different numbers, q_0 is a node of G, and Q is a nonempty set of the graph G nodes.

Let $I=(G,q_0,Q)$ be an A-source over the alphabet E_k. An *I-trace* in the graph G is an arbitrary sequence $\tau=v_1,d_1,\ldots,v_m,d_m,v_{m+1}$ of nodes and edges of G such that $v_1=q_0$, $v_{m+1}\in Q$, and v_i is the initial and v_{i+1} is the terminal node of the edge

d_i for $i = 1, \ldots, m$. Now we define a word word(τ) from $(E_k)^*$ in the following way: if $m = 0$ then word(τ) $= \lambda$. Let $m > 0$, and let δ_j be the number assigned to the edge d_j, $j = 1, \ldots, m$. Then word(τ) $= \delta_1 \ldots \delta_m$. We can extend the notation word(τ) to an arbitrary directed path τ in the graph G. We denote by $\Xi(I)$ the set of all I-traces in G. Let $E(I) = \{\text{word}(\tau) : \tau \in \Xi(I)\}$. We will say that the source I *generates* the language $E(I)$. It is well known that $E(I)$ is a regular language.

The A-source I will be called *reduced* if for each node of G there exists an I-trace which contains this node. Further we will assume that a considered regular language is nonempty and it is given by a reduced A-source which generates this language (it is known that such A-source exists).

1.6.2.3 Types of Reduced A-Sources

Let $I = (G, q_0, Q)$ be a reduced A-source over the alphabet E_k. A directed path in the graph G will be called a *path of the source I*. A path of the source I will be called a *cycle of the source I* if there is at least one edge in this path, and the first node of this path is equal to the last node of this path. A cycle of the source I will be called *elementary* if nodes of this cycle, with the exception of the last node, are pairwise different. Sometimes, cycles of the source I will be considered as subgraphs of the graph G.

The source I will be called *simple* if each two different (as subgraphs) elementary cycles of the source I do not have common nodes. Let I be a simple source and τ be an I-trace. The number of different (as subgraphs) elementary cycles of the source I, which have common nodes with τ, will be denoted by $cl(\tau)$ and will be called the *cyclic length of the path* τ. The value $cl(I) = \max\{cl(\tau) : \tau \in \Xi(I)\}$ will be called the *cyclic length of the source I*.

Let I be a simple source, C be an elementary cycle of the source I, and v be a node of the cycle C. Beginning with the node v, the cycle C generates an infinite periodic word over the alphabet E_k. This word will be denoted by $W(I, C, v)$. We denote by $r(I, C, v)$ the minimum period of the word $W(I, C, v)$. We denote by $l(C)$ the number of nodes in the elementary cycle C (the *length* of C).

The source I will be called *dependent* if there exist two different (as subgraphs) elementary cycles C_1 and C_2 of the source I, nodes v_1 and v_2 of the cycles C_1 and C_2 respectively, and a path π of the source I from v_1 to v_2 which satisfy the following conditions: $W(I, C_1, v_1) = W(I, C_2, v_2)$ and the length of the path π is a number divisible by $r(I, C_1, v_1)$. If the source I is not dependent then it will be called *independent*.

The source I will be called *strongly dependent* if in I there exist pairwise different (as subgraphs) elementary cycles $C_1, \ldots, C_m$, $(m \geq 1)$, and pairwise different (as subgraphs) elementary cycles $B_1, \ldots, B_m$, $D_1, \ldots, D_m$, vertices $v_0, \ldots, v_{m+1}$, $u_1, \ldots, v_{m+1}$, $w_0, \ldots, w_m$ and paths $\tau_0, \ldots, \tau_m$, $\pi_0, \ldots, \pi_m$, $\gamma_1, \ldots, \gamma_m$ which satisfy the following conditions:

a) $v_0 = w_0 = q_0$, $v_{m+1} \in Q$, $u_{m+1} \in Q$;

b) for $i = 1, \ldots, m$, the node v_i belongs to the cycle C_i, the node u_i belongs to the cycle B_i, and the node w_i belongs to the cycle D_i;

c) τ_i is a path from v_i to v_{i+1}, and π_i is a path from w_i to u_{i+1}, $i = 0, \ldots, m$;
d) γ_i is a path from u_i to w_i, $i = 1, \ldots, m$;
e) $W(I, C_i, v_i) = W(I, B_i, u_i) = W(I, D_i, w_i)$ for $i = 1, \ldots, m$;
f) word(τ_i) = word(π_i) for $i = 0, \ldots, m$;
g) for $i = 1, \ldots, m$, the length of the path γ_i is a number divisible by $l(C_i)$.

One can show that if the source I is strongly dependent then the source I is dependent.

1.6.2.4 Main Result

In the following theorem, the behavior of functions $H_{\mathscr{L}}$ and $P_{\mathscr{L}}$ is considered.

Theorem 1.113. *Let $\mathscr{L}$ be a nonempty regular language and I be a reduced A-source which generates the language $\mathscr{L}$. Then*

a) if I is an independent simple source and $cl(I) \leq 1$ then there exists a constant $c_1 \in \omega \setminus \{0\}$ such that for any $n \in \omega \setminus \{0\}$ the following inequalities hold:

$$P_{\mathscr{L}}(n) \leq H_{\mathscr{L}}(n) \leq c_1 \,;$$

b) if I is an independent simple source and $cl(I) \geq 2$ then there exist constants $c_1, c_2, c_3, c_4, c_5 \in \omega \setminus \{0\}$ such that for any $n \in \omega \setminus \{0\}$ the following inequalities hold:

$$\frac{\log_2 n}{c_1} - c_2 \leq H_{\mathscr{L}}(n) \leq c_3 \log_2 n + c_4 \text{ and } P_{\mathscr{L}}(n) \leq c_5 \,;$$

c) if I is a dependent simple source which is not strongly dependent then there exist constants $c_1, c_2, c_3 \in \omega \setminus \{0\}$ such that for any $n \in \omega \setminus \{0\}$ the following inequalities hold:

$$\frac{n}{c_1} - c_2 \leq H_{\mathscr{L}}(n) \leq n \text{ and } P_{\mathscr{L}}(n) \leq c_3 \,;$$

d) if I is a strongly dependent simple source or I is not a simple source then there exist constants $c_1, c_2 \in \omega \setminus \{0\}$ such that for any $n \in \omega \setminus \{0\}$ the following inequalities hold:

$$\frac{n}{c_1} - c_2 \leq P_{\mathscr{L}}(n) \leq H_{\mathscr{L}}(n) \leq n \,.$$

Example 1.114. Let I be the source depicted in Fig. 1.11 and $\mathscr{L}$ be the regular language which is generated by I. The source I is an independent simple A-source with $cl(I) = 2$. One can show that $H_{\mathscr{L}}(n) = \lceil \log_2 n \rceil$ for any $n \in \omega \setminus \{0\}$, $P_{\mathscr{L}}(n) = n - 1$ for $n = 1, 2$ and $P_{\mathscr{L}}(n) = 2$ for any $n \in \omega \setminus \{0\}, n \geq 3$.

Similar results for languages generated by some types of linear grammars and context-free grammars were obtained by Dudina and Knyazev [18, 28, 29].

1.7 Conclusions

In this chapter, we tried to achieve two goals: (i) describe main areas of TT and some results of TT (it was impossible for us to consider all results obtained during more than 50 years), and (ii) describe our own view on areas of TT in which we worked.

We considered the most known ways of TT development: a variety of tools for the study of tests, and main areas of applications of TT: control and diagnosis of faults, and pattern recognition (prediction).

We studied tests, decision trees, and decision rule systems from a common point of view which is close both to TT and RS. We considered also less known and developed in lesser degree areas of TT. We studied decision trees and decision rule systems over infinite or large finite sets of attributes. This is an unusual area for TT, as well as two more areas of applications – discrete optimization and recognition of words from formal languages.

The most part of results in TT were published in Russian with their own terminology. We tried to unify the terminology and to move it closer to papers related to data analysis and written in English. We hope that it makes the results of TT accessible for wider groups of researchers.

References

1. Alkhalid, A., Chikalov, I., Hussain, S., Moshkov, M.: Extensions of Dynamic Programming as a New Tool for Decision Tree Optimization. In: Ramanna, S., Howlett, R.J., Jain, L.C. (eds.) Emerging Paradigms in Machine Learning and Applications. SIST, vol. 13, pp. 17–36. Springer, Heidelberg (2012)
2. Alkhalid, A., Chikalov, I., Moshkov, M.: On Algorithm for Building of Optimal α-Decision Trees. In: Szczuka, M., Kryszkiewicz, M., Ramanna, S., Jensen, R., Hu, Q. (eds.) RSCTC 2010. LNCS, vol. 6086, pp. 438–445. Springer, Heidelberg (2010)
3. Amin, T., Chikalov, I., Moshkov, M., Zielosko, B.: Dynamic programming algorithm for optimization of β-decision rules. In: Szczuka, M., Czaja, L., Skowron, A., Kacprzak, M. (eds.) Proc. of Int'l Workshop Concurrency, Specification and Programming, CS&P 2011, Pułtusk, Poland, pp. 10–16. Białystok University of Technology (2011)
4. Amin, T., Chikalov, I., Moshkov, M., Zielosko, B.: Dynamic programming approach for exact decision rule optimization. In: Skowron, A., Suraj, Z. (eds.) Special Volume Dedicated to the Memory of Professor Zdzislaw Pawlak. ISRL, Springer (to appear, 2012)
5. Armstrong, D.B.: On finding a nearly minimal set of fault detection tests for combinatorial logic nets. IEEE Trans. Electron. Comput. EC-15, 66–73 (1966)
6. Baskalova, L.V., Zhuravlev, Y.I.: A model of recognition algorithms with representative samples and systems of supporting sets. USSR Comput. Maths. and Mathematical Phys. 21(5), 189–199 (1981)
7. Borodina, Y.V.: Synthesis of easily-tested circuits in the case of single-type constant malfunctions at the element outputs. In: Proc. of the Ninth International Seminar on Discrete Mathematics and its Applications, pp. 64–65. MSU Publishers, Moscow (2007) (in Russian)
8. Borodina, Y.V.: Synthesis of easily-tested circuits in the case of single-type constant malfunctions at the element outputs. Moscow University Computational Mathematics and Cybernetics 32, 42–46 (2008)
9. Borodina, Y.V.: Synthesis of easily-tested circuits in the case of single-type constant malfunctions at the element outputs. Abstract of dissertation for the degree of candidate of physical and mathematical sciences (2008) (in Russian)
10. Chegis, I.A., Yablonskii, S.V.: Logical methods of control of work of electric schemes. Trudy Mat. Inst. Steklov. 51, 270–360 (1958) (in Russian)
11. Chikalov, I.: Algorithm for constructing of decision trees with minimal average depth. In: Proc. of the Eighth Int'l Conference on Information Processing and Management of Uncertainty in Knowledge-Based Systems, Madrid, Spain, vol. 1, pp. 376–379 (2000)

12. Chikalov, I.V.: On Algorithm for Constructing of Decision Trees with Minimal Number of Nodes. In: Ziarko, W.P., Yao, Y. (eds.) RSCTC 2000. LNCS (LNAI), vol. 2005, pp. 139–143. Springer, Heidelberg (2001)
13. Chikalov, I.: Average Time Complexity of Decision Trees. ISRL, vol. 21. Springer, Heidelberg (2011)
14. Chikalov, I., Hussain, S., Moshkov, M.: Relationships for cost and uncertainty of decision trees. In: Skowron, A., Suraj, Z. (eds.) Special Volume Dedicated to the Memory of Professor Zdzisław Pawlak. ISRL. Springer (to appear, 2012)
15. Chikalov, I.V., Moshkov, M.J., Zelentsova, M.S.: On Optimization of Decision Trees. In: Peters, J.F., Skowron, A. (eds.) Transactions on Rough Sets IV. LNCS, vol. 3700, pp. 18–36. Springer, Heidelberg (2005)
16. Dmitriyev, A.N., Zhuravlev, Y.I., Krendelev, F.P.: On mathematical principles for classification of objects and phenomena. Diskret. Analiz 7, 3–15 (1966) (in Russian)
17. Dmitriyev, A.N., Zhuravlev, Y.I., Krendelev, F.P.: On a principle of classification and prediction of geological objects and phenomena. Geol. Geofiz. 5, 50–64 (1968) (in Russian)
18. Dudina, J.V., Knyazev, A.N.: On complexity of recognition of words from languages generated by context-free grammars with one nonterminal symbol. In: Vestnik of Lobachevsky State University of Nizhny Novgorod. Mathematical Simulation and Optimal Control, vol. 2, pp. 214–223 (1998) (in Russian)
19. Dyukova, E.V., Zhuravlev, Y.I.: Discrete analysis of feature descriptions in recognition problems of high dimensionality. Computational Mathematics and Mathematical Physics 40, 1214–1227 (2000)
20. Eldred, R.D.: Test routines based on symbolic logic statements. J. ACM 6(1), 33–36 (1959)
21. Feige, U.: A threshold of $\ln n$ for approximating set cover (preliminary version). In: Proc. of the Twenty-Eighth Annual ACM Symposium on Theory of Computing, pp. 314–318. ACM, New York (1996)
22. Goldman, R.S., Chipulis, V.P.: Diagnosis of iteration-free combinatorial circuits. Diskret. Analiz 14, 3–15 (1969) (in Russian)
23. Goldman, S.A., Kearns, M.J.: On the complexity of teaching. In: Warmuth, M.K., Valiant, L.G. (eds.) Proc. of the Fourth Annual Workshop on Computational Learning Theory, COLT 1991, Santa Cruz, California, USA, pp. 303–314. Morgan Kaufmann (1991)
24. Hegedüs, T.: Generalized teaching dimensions and the query complexity of learning. In: Maass, W. (ed.) Proc. of the Eigth Annual Conference on Computational Learning Theory, COLT 1995, Santa Cruz, California, USA, pp. 108–117. ACM, New York (1995)
25. Hellerstein, L., Pillaipakkamnatt, K., Raghavan, V., Wilkins, D.: How many queries are needed to learn. J. ACM 43, 840–862 (1996)
26. Johnson, D.S.: Approximation algorithms for combinatorial problems. Journal of Computer and System Sciences 9, 256–278 (1974)
27. Karavai, M.F.: Diagnosis of tree-like circuits in arbitrary basis. Automation and Telemechanics 1, 173–181 (1973) (in Russian)
28. Knyazev, A.: On Recognition of Words from Languages Generated by Linear Grammars with one Nonterminal Symbol. In: Polkowski, L., Skowron, A. (eds.) RSCTC 1998. LNCS (LNAI), vol. 1424, pp. 111–114. Springer, Heidelberg (1998)
29. Knyazev, A.: On recognition of words from languages generated by context-free grammars with one nonterminal symbol. In: Proc. of the Eighth Int'l Conference on Information Processing and Management of Uncertainty in Knowledge-Based Systems, Madrid, Spain, vol. 1, pp. 1945–1948 (2000)

30. Konstantinov, R.M., Koroleva, Z.E., Kudryavtsev, V.B.: Combinatory logic approach to problems of predicting ore yields. Problemy Kibernet. 31, 25–41 (1976) (in Russian)
31. Korshunov, A.D.: The length of minimum tests for rectangular tables. I. Cybernetics and Systems Analysis 6, 723–733 (1970)
32. Kospanov, E.S.: An algorithm for the construction of sufficiently simple tests. Diskret. Analiz 1, 43–47 (1966) (in Russian)
33. Kudryavtsev, V.B., Andreev, A.E.: Test recognition. Journal of Mathematical Sciences 169(4), 457–480 (2010)
34. Laskowski, M.C.: Vapnik-Chervonenkis classes of definable sets. J. London Math. Society 45, 377–384 (1992)
35. Madatyan, K.A.: On complete tests for contact circuits without repetitions. Problemy Kibernet. 23, 103–118 (1970) (in Russian)
36. Madatyan, K.A.: Construction of single test for contact circuits. In: Collection of Works on Mathematical Cybernetics, pp. 77–86. Computer Centre of USSR Academy of Sciences, Moscow (1981) (in Russian)
37. Moore, E.F.: Gedanken-experiments on sequential machines. In: Automata Studies. Annals of Mathematical Studies, vol. 34, pp. 129–153. Princeton University Press, Princeton (1956)
38. Moshkov, M.J.: Conditional tests. Problemy Kibernet. 40, 131–170 (1983) (in Russian)
39. Moshkov, M.J.: Decision trees with quasilinear checks. Trudy IM SO RAN 27, 108–141 (1994) (in Russian)
40. Moshkov, M.J.: Complexity of deterministic and nondeterministic decision trees for regular language word recognition. In: Bozapalidis, S. (ed.) Proc. of the 3rd International Conference Developments in Language Theory, DLT 1997, Thessaloniki, Greece, pp. 343–349. Aristotle University of Thessaloniki (1997)
41. Moshkov, M.J.: Diagnosis of constant faults in circuits. In: Mathematical Problems of Cybernetics, vol. 9, pp. 79–100. Nauka Publishers, Moscow (2000) (in Russian)
42. Moshkov, M.J.: Time Complexity of Decision Trees. In: Peters, J.F., Skowron, A. (eds.) Transactions on Rough Sets III. LNCS, vol. 3400, pp. 244–459. Springer, Heidelberg (2005)
43. Moshkov, M.J., Chikalov, I.: On algorithm for constructing of decision trees with minimal depth. Fundam. Inform. 41(3), 295–299 (2000)
44. Moshkov, M.J., Piliszczuk, M., Zielosko, B.: Partial Covers, Reducts and Decision Rules in Rough Sets – Theory and Applications. SCI, vol. 145. Springer, Heidelberg (2008)
45. Moshkov, M.J., Zielosko, B.: Combinatorial Machine Learning – A Rough Set Approach. SCI, vol. 360. Springer, Heidelberg (2011)
46. Moshkova, A.: On Diagnosis of Retaining Faults in Circuits. In: Polkowski, L., Skowron, A. (eds.) RSCTC 1998. LNCS (LNAI), vol. 1424, pp. 513–516. Springer, Heidelberg (1998)
47. Moshkova, A.: On time complexity of retaining fault diagnosis in circuits. In: Proc. of the Eighth Int'l Conference on Information Processing and Management of Uncertainty in Knowledge-Based Systems, Madrid, Spain, vol. 1, pp. 372–375 (2000)
48. Nigmatullin, R.G.: The fastest descent method for covering problems. In: Proc. Questions of Precision and Efficiency of Computer Algorithms, Kiev, USSR, vol. 5, pp. 116–126 (1969) (in Russian)
49. Noskov, V.: On dead-end and minimal tests for a certain class of tables. Diskret. Analiz 12, 27–49 (1968) (in Russian)
50. Noskov, V.N.: Diagnostic tests for logic diagram inputs. Diskret. Analiz 26, 72–83 (1974) (in Russian)

51. Noskov, V.N.: Complexity of tests checking the operation of logic diagram inputs. Diskret. Analiz 27, 23–51 (1975) (in Russian)
52. Noskov, V.N.: On length of minimal single diagnostic tests checking the operation of logic diagram inputs. Metody Diskret. Analiz 32, 40–51 (1978) (in Russian)
53. Noskov, V.N., Slepyan, V.A.: Number of dead-end tests for a certain class of tables. Cybernetics and Systems Analysis 8, 64–71 (1972)
54. Preparata, F.P.: An estimate of the length of diagnostics tests. IEEE Transactions on Reliability R-18(3), 131–136 (1969)
55. Red'kin, N.P.: Complete detection tests for switching circuits. Metody Diskret. Analiz 39, 80–87 (1983) (in Russian)
56. Red'kin, N.P.: Verifying tests for closed and broken circuits. Metody Diskret. Analiz 40, 87–99 (1983) (in Russian)
57. Red'kin, N.P.: On complete checking tests for circuits of functional elements. Moscow State University Bulletin, Series 1 – Mathematics and Mechanics (1), 72–74 (1986) (in Russian)
58. Red'kin, N.P.: Single checking tests for schemes with inverse errors of elements. Mathematical Problems of Cybernetics 12, 217–230 (2003) (in Russian)
59. Roth, J.P.: Diagnosis of automata failures: a calculus and a method. IBM J. Res. Develop. 10, 278–291 (1966)
60. Shevtchenko, V.I.: On the depth of conditional tests for controlling "negation" type faults in circuits of functional gates. Sibirsk. Zh. Issled. Oper. 1(1), 63–74 (1994) (in Russian)
61. Shevtchenko, V.I.: On complexity of fault diagnosis of circuits of functional elements. In: Lupanov, O., Chashkin, A. (eds.) Collection of Lectures of Youth Scientific Schools on Discrete Mathematics and its Applications, vol. 2, pp. 111–123. MSU Publishers, Moscow (2001) (in Russian)
62. Slavík, P.: A tight analysis of the greedy algorithm for set cover. In: Proc. of the Twenty-Eighth Annual ACM Symposium on Theory of Computing, pp. 435–441. ACM, New York (1996)
63. Slavík, P.: Approximation algorithms for set cover and related problems. Ph.D. thesis, State University of New York at Buffalo (1998)
64. Slepyan, V.: The parameters of the distribution of dead-end tests and the information weights of columns in binary tables. Diskret. Analiz 14, 28–43 (1969) (in Russian)
65. Slepyan, V.: The length of minimal test for a certain class of tables. Diskret. Analiz 23, 59–71 (1973) (in Russian)
66. Soloviev, N.: On certain property of tables with dead-end tests of equal length. Diskret. Analiz 12, 91–95 (1968) (in Russian)
67. Soloviev, N.: On tables containing trivial dead-end tests. Diskret. Analiz 12, 96–114 (1968) (in Russian)
68. Soloviev, N.A.: Tests (Theory, Construction, Applications), Nauka, Novosibirsk (1978) (in Russian)
69. Vaintsvaig, M.N.: Pattern recognition learning algorithm Kora. In: Pattern Recognition Learning Algorithms, pp. 110–116. Sovetskoe Radio, Moscow (1973) (in Russian)
70. Vapnik, V.N., Chervonenkis, A.Y.: On the uniform convergence of relative frequencies of events to their probabilities. Theory of Probability and its Applications 16, 264–280 (1971)
71. Vasilevskii, M.P.: Failure diagnosis of automata. Cybernetics and Systems Analysis 9(4), 653–665 (1973)
72. Vasilevskii, M.P.: Concerning the decoding of automata. Cybernetics and Systems Analysis 10(2), 213–218 (1974)

73. Yablonskii, S.V.: On the construction of dead-end multiple experiments for automata. Trudy Mat. Inst. Steklov. 83, 263–272 (1973) (in Russian)
74. Yablonskii, S.V.: Some problems of reliability and diagnosis in control systems. Mathematical Problems of Cybernetics 1, 5–25 (1988) (in Russian)
75. Yablonskii, S.V., Chegis, I.A.: On tests for electric circuits. Uspekhi Mat. Nauk 10, 182–184 (1955) (in Russian)
76. Zhuravlev, Y.I.: A class of partial Boolean functions. Diskret. Analiz 2, 23–27 (1964) (in Russian)
77. Zhuravlev, Y.I., Petrov, I.B., Ryazanov, V.V.: Discrete methods of diagnosis and analysis of medical information. In: Medicine in the Mirror of Informatics, Nauka Moscow, pp. 113–123 (2008) (in Russian)
78. Zielosko, B., Moshkov, M., Chikalov, I.: Decision rule optimization on the basis of dynamic programming methods. Vestnik of Lobachevsky State University of Nizhny Novgorod 6, 195–200 (2010) (in Russian)

Sergei V. Yablonskii (1924 -1998)

Sergei V. Yablonskii is a Soviet and Russian mathematician, corresponding member of USSR Academy of Sciences (since 1968), one of the founders of the national school of mathematical cybernetics, and mathematical theory of control systems.

Yablonskii was born on December 6, 1924 in Moscow. His father was a scientist graduated from Moscow State University and then teaching at Moscow Institute of Oil. Sergei loved reading of science fiction books and in the secondary school organized a kind of scientific society with his classmates, where they were discussing different problems and were publishing reports. His mathematical talent manifested itself at school, and in 1940 he won the 6th Moscow Mathematical Olympiad. Next year, he enrolled in faculty of Mechanics and Mathematics of Moscow State University. The same year the USSR entered World War II, and in autumn 1942 after the first course Yablonskii went to the front eighteen years old. He was awarded five orders and several medals for combat service.

Yablonskii returned to MSU in 1945 and in 1950 graduated from the University. In 1950 he published his first scientific paper, "On convergent sequences of continuous functions" in the Bulletin of Moscow State University. In 1950 Yablonskii entered post graduate school of Mechanics and Mathematics Faculty of Moscow State University. He concentrated his research on problems of expressibility in mathematical logic. His work showed that some problems stated in mathematical logic have more adequate description by means of discrete-valued functions. In 1953 he became candidate of science. His thesis "Questions of functional completeness in k-valued calculus" contributes to the development of mathematical logic theory and solves several problems (in particular, the problem of completeness in three-valued logic). Since 1953, Yablonskii worked at the Institute of Applied Mathematics of Soviet Academy of Sciences. He continued his studies in the field of discrete-valued functions, and in 1958 published a review article, "Functional constructions in k-valued logic" in Proceedings of the Steklov Institute of Mathematics, which successfully systematized results in this area. This paper has played an important role in the formation of discrete mathematics and mathematical cybernetics, and for many years been a major textbook on the theory of discrete functions.

At the same time Yablonskii was actively investigating problems related to the synthesis of logic devices. In collaboration with I. Chegis, in 1958 he published a paper "Logical methods of control of work of electric schemes" in Proceedings of the Steklov Institute of Mathematics. The paper presents a new look to the problem of constructing tests and boosts development of combinatorial-logical methods in circuit reliability theory and pattern recognition. Yablonskii actively supported an emerging area of science, cybernetics, and analyzed its mathematical aspects. His view is presented in the article "Basic concepts of cybernetics", published in 1959 in the book "Problems of Cybernetics". The article highlighted and mathematically formalized the concept of the control system and pointed out the problems and directions of development of the theory of control systems. Along with Alexei A. Lyapunov, in the 50s and 60s Yablonskii chaired a famous seminar on cybernetics. He took an active part in organizing of a periodical "Problems of Cybernetics", which started publishing in 1958 by Lyapunov.

Research School of Yablonskii, established in Moscow, went far beyond it. Yablonskii took part in organizing and holding of the first USSR-wide conferences on problems of theoretical cybernetics, and then for many years was the permanent chairman of the organizing committee. He contributed to emergence and growth of research teams in Nizhny Novgorod, Novosibirsk, Kazan, Saratov, Irkutsk and other cities. Yablonskii took part in organizing S. Banach International Mathematical Center in Warsaw. He has been a board member of the center and participated in organizing semesters of discrete mathematics.

In 1958, Yablonskii created department of cybernetics at the Institute of Applied Mathematics and headed it for more than 40 years. During this period he was engaged in research in complexity of algorithms for minimization of circuits. The received results explaining the difficulties in constructing minimal circuits are included in his doctoral thesis "On some mathematical problems of the theory of control systems", which he defended in 1962. In 1966, Yablonskii (with Zhuravlev and Lupanov) was awarded the Lenin Prize for his work on the theory of control systems. In 1968 he was elected a corresponding member of USSR Academy of Sciences. Yablonskii for a long time has been productively working at leading positions in the Department of Mathematics of the USSR Academy of Science. He was also a founding member of the Academy of Cryptography.

Since 1954, Yablonskii has taught at the Moscow State University. His lectures and seminars on mathematical problems of cybernetics at Mechanics and Mathematics Faculty form a basis of contemporary university courses on this and related disciplines. He prepared 10 doctors and 25 candidates of science. In 1971 Yablonskii created chair of automata theory and mathematical logic (later renamed the Department of Mathematical Cybernetics) at the Faculty of Computational Mathematics and Cybernetics, and headed it until his death. He was a member of the editorial board of "Mathematical Encyclopaedia" and a number of scientific journals.

Yablonskii's work on functional constructions in many-valued logics continued Post's work on two-valued logic. He established a criterion of completeness for 3-valued logic in terms of precomplete classes and found a number of maximal classes in other multi-valued logics. These results formed the basis for Rosenberg's integral result, a completeness criterion in any finite-valued logic.

Yablonskii first formulated and proved a mathematical statement justifying the hypothesis of the inevitability of a large enumeration when constructing optimal control systems and a theorem that demonstrates algorithmic difficulties of synthesis of minimal combinatorial circuits. In the synthesis of control systems he described and studied a continuous family of classes of Boolean functions that allow a simpler implementation than most of the Boolean functions. He contributed greatly to the reliability theory, finding that it is possible to build reliable circuits (in various senses) without a significant complication of circuits. He constructed a general theory of tests for the control systems.

The works of Yablonskii on difficulties in algorithmic synthesis of minimal circuits, on the completeness of functional systems, reliability and diagnosis of control systems still have fundamental importance.

Part II

Rough Sets

2

Rough Sets

2.1 Introduction

Rough set theory, proposed by Professor Zdzisław Pawlak in 1982 [163, 165, 166, 169], can be seen as a new mathematical approach to dealing with imperfect knowledge, in particular with vague concepts. The rough set philosophy is founded on the assumption that with every object of the universe of discourse we associate some information (data, knowledge). For example, if objects are patients suffering from a certain disease, symptoms of the disease form information about patients. Objects characterized by the same information are indiscernible (similar) in view of the available information about them. The indiscernibility relation generated in this way is the mathematical basis of rough set theory. This understanding of indiscernibility is related to the idea of Gottfried Wilhelm Leibniz that objects are indiscernible if and only if all available functionals take on identical values (Leibniz's Law of Indiscernibility: The Identity of Indiscernibles) [4, 97]. However, in the rough set approach, indiscernibility is defined relative to a given set of functionals (attributes).

Any set of all indiscernible (similar) objects is called an elementary set, and forms a basic granule (atom) of knowledge about the universe. Any union of elementary sets is referred to as a crisp (precise) set[1]. A set which is not crisp is called rough (imprecise, vague).

Consequently, each rough set has boundary region cases, i.e., objects which cannot with certainty be classified either as members of the set or of its complement. Obviously, crisp sets have no boundary region elements at all. This means that boundary region cases cannot be properly classified by employing available knowledge.

Thus, the assumption that objects can be "seen" only through the information available about them leads to the view that knowledge has a granular structure. Due to the granularity of knowledge, some objects of interest cannot be discerned and appear as the same (or similar). As a consequence, vague concepts, in contrast to precise concepts, cannot be characterized in terms of information about their elements.

[1] This approach is generalized when one considers inductive extensions of approximations from samples of objects (see, e.g., [246]).

I. Chikalov et al.: Three Approaches to Data Analysis, ISRL 41, pp. 69–135.
springerlink.com

Therefore, in the proposed approach, we assume that any vague concept is replaced by a pair of precise concepts – called the lower and the upper approximation of the vague concept. The lower approximation consists of all objects which surely belong to the concept and the upper approximation contains all objects which possibly belong to the concept. The difference between the upper and the lower approximation constitutes the boundary region of the vague concept. These approximations are two basic operations in rough set theory. Note, that the boundary region is defined relative to a subjective knowledge given by a set of attributes or/and sample of objects. Such a boundary region is crisp. However, when some attributes are deleted, new attributes are added or a given sample is updated the boundary region is changing. One could ask about a boundary region independent of such subjective knowledge but then, in the discussed framework, we do not have a possibility to define such region as a crisp set. This property is related to the higher order vagueness discussed in philosophy.

Hence, rough set theory expresses vagueness not by means of membership, but by employing a boundary region of a set. If the boundary region of a set is empty, it means that the set is crisp, otherwise the set is rough (inexact). A nonempty boundary region of a set means that our knowledge about the set is not sufficient to define the set precisely.

Rough set theory it is not an alternative to but rather is embedded in classical set theory. Rough set theory can be viewed as a specific implementation of Frege's idea of vagueness [48], i.e., imprecision in this approach is expressed by a boundary region of a set.

Rough set theory has attracted worldwide attention of many researchers and practitioners, who have contributed essentially to its development and applications. Rough set theory overlaps with many other theories. Despite this, rough set theory may be considered as an independent discipline in its own right. The rough set approach seems to be of fundamental importance in artificial intelligence and cognitive sciences, especially in research areas such as machine learning, intelligent systems, inductive reasoning, pattern recognition, mereology, image processing, signal analysis, knowledge discovery, decision analysis, and expert systems. The main advantage of rough set theory in data analysis is that it does not need any preliminary or additional information about data like probability distributions in statistics, basic probability assignments in Dempster–Shafer theory, a grade of membership or the value of possibility in fuzzy set theory (see, e.g., [44] where some combinations of rough sets with non-parametric statistics are studied). One can observe the following about the rough set approach:

- introduction of efficient algorithms for finding hidden patterns in data,
- determination of optimal sets of data (data reduction),
- evaluation of the significance of data,
- generation of sets of decision rules from data,
- easy-to-understand formulation,
- straightforward interpretation of obtained results,
- suitability of many of its algorithms for parallel processing.

The basic ideas of rough set theory and its extensions as well as many interesting applications can be found in a number of books (see, e.g., [34, 37, 41, 44, 69, 80, 91, 92, 105, 145, 153, 154, 169, 196, 200, 203, 204, 224, 266, 319, 85, 118, 32, 76, 176, 36, 23, 86, 152, 121, 199, 108]), issues of the Transactions on Rough Sets [185, 183, 178, 179, 180, 184, 187, 181, 188, 193, 182, 190, 186, 189], special issues of other journals (see, e.g., [31, 101, 177, 151, 235, 268, 323, 324, 81, 147, 35, 236]), proceedings of international conferences (see, e.g., [2, 78, 104, 202, 234, 253, 262, 263, 289, 291, 292, 303, 322, 326, 59, 304, 3, 309, 95, 33, 302, 305, 221, 286, 316, 96, 310]), tutorials (see, e.g., [90, 175, 174, 173]). For more information on the bibliography on rough sets and rough set based software systems one can also visit web pages[2].

In the development of rough set theory and applications, one can distinguish three main stages. At the beginning, the researchers were concentrated on descriptive properties such as reducts of information systems preserving indiscernibility relations or description of concepts or classifications. Next, they moved to applications of rough sets in machine learning, pattern recognition, and data mining. After gaining some experiences, they developed foundations for inductive reasoning leading to, e.g., inducing classifiers. While the first period was based on the assumption that objects are perceived by means of partial information represented by attributes, in the second period it was also used the assumption that information about the approximated concepts is partial too. Approximation spaces and strategies of searching for relevant approximation spaces were recognized as the basic tools for rough sets. Important achievements both in theory and applications were obtained using Boolean reasoning and approximate Boolean reasoning applied, e.g., in searching for relevant features, discretization, symbolic value grouping, or, in more general sense, in searching for relevant approximation spaces. Nowadays, we observe that a new period is emerging in which two new important topics are investigated: (i) strategies for discovering relevant (complex) contexts of analyzed objects or granules, what is strongly related to information granulation process and granular computing, and (ii) interactive computations on granules. Both directions are aiming at developing tools for approximation of complex vague concepts such as behavioral patterns or adaptive strategies making it possible to achieve the satisfactory qualities of realized interactive computations.

This chapter presents this development from rudiments of rough sets to challenges. We begin with a short discussion on vague concepts (see Section 2.2). Next, we recall the basic concepts of rough set theory (see Section 2.3). Some extensions of the rough set approach are outlined in Section 2.4. In Section 2.5, we discuss the relationship of the rough set approach with inductive reasoning. In particular, we present the rough set approach to inducing rough set based classifiers and inducing relevant approximation spaces. We also discuss shortly the relationship of the rough set approach and the higher order vagueness. Section 2.6 includes some remarks on relationships of information granulation and rough sets. In Section 2.7, we outline the rough set approach to ontology approximation. The rough set approach based

[2] `www.roughsets.org`

on combination of rough sets and Boolean reasoning to scalability in data mining is discussed in Section 2.8.

Some comments on relationships of rough sets and logic are included in Section 2.9.

Finally, we discuss some challenging issues for rough sets (see Section 2.10). We propose Interactive Rough Granular Computing (IRGC) as a framework making it possible to search for solutions of problems related to inducing of relevant contexts, process mining and perception based copmputing (PBC).

2.2 Vague Concepts

Mathematics requires that all mathematical notions (including set) must be exact, otherwise precise reasoning would be impossible. However, philosophers [87, 88, 216, 220] and recently computer scientists [109, 140, 142, 232] as well as other researchers have become interested in *vague* (imprecise) concepts.

In classical set theory, a set is uniquely determined by its elements. In other words, this means that every element must be uniquely classified as belonging to the set or not. That is to say the notion of a set is a *crisp* (precise) one. For example, the set of odd numbers is crisp because every number is either odd or even.

In contrast to odd numbers, the notion of a beautiful painting is vague, because we are unable to classify uniquely all paintings into two classes: *beautiful* and not *beautiful*. Some paintings cannot be decided whether they are beautiful or not and thus they remain in the doubtful area. Thus, *beauty* is not a precise but a vague concept.

Almost all concepts we are using in natural language are vague. Therefore, common sense reasoning based on natural language must be based on vague concepts and not on classical logic. Interesting discussion of this issue can be found in [216].

The idea of vagueness can be traced back to the ancient Greek philosopher Eubulides of Megara (ca. 400BC) who first formulated so called "sorites" (heap) and "falakros" (bald man) paradoxes (see, e.g., [87, 88]). The bald man paradox goes as follows: suppose a man has 100,000 hairs on his head. Removing one hair from his head surely cannot make him bald. Repeating this step we arrive at the conclusion the a man without any hair is not bald. Similar reasoning can be applied to a hip of stones.

Vagueness is usually associated with the boundary region approach (i.e., existence of objects which cannot be uniquely classified relative to a set or its complement) which was first formulated in 1893 by the father of modern logic, German logician, Gottlob Frege (1848-1925) (see [48]).

According to Frege the concept must have a sharp boundary. To the concept without a sharp boundary there would correspond an area that would not have any sharp boundary–line all around. It means that mathematics must use crisp, not vague concepts, otherwise it would be impossible to reason precisely.

Summing up, vagueness is

- not allowed in mathematics;
- interesting for philosophy;

- a nettlesome problem for natural language, cognitive science, artificial intelligence, machine learning, philosophy, and computer science.

2.3 Rudiments of Rough Sets

This section briefly delineates basic concepts in rough set theory.

2.3.1 Indiscernibility and Approximation

The starting point of rough set theory is the indiscernibility relation, which is generated by information about objects of interest (see Sect. 2.1). The indiscernibility relation expresses the fact that due to a lack of information (or knowledge) we are unable to discern some objects employing available information (or knowledge).

This means that, in general, we are unable to deal with each particular object but we have to consider granules (clusters) of indiscernible objects as a fundamental basis for our theory.

From a practical point of view, it is better to define basic concepts of this theory in terms of data. Therefore we will start our considerations from a data set called an *information system*. An information system is a data table containing rows labeled by objects of interest, columns labeled by attributes and entries of the table are attribute values. For example, a data table can describe a set of patients in a hospital. The patients can be characterized by some attributes, like *age, sex, blood pressure, body temperature*, etc. With every attribute a set of its values is associated, e.g., values of the attribute *age* can be *young, middle*, and *old*. Attribute values can be also numerical. In data analysis the basic problem we are interested in is to find patterns in data, i.e., to find a relationship between some sets of attributes, e.g., we might be interested whether *blood pressure* depends on *age and sex*.

Suppose we are given a pair $\mathbb{A} = (U, A)$ of non-empty, finite sets U and A, where U is the *universe* of *objects*, and A – a set consisting of *attributes*, i.e. functions $a : U \longrightarrow V_a$, where V_a is the set of values of attribute a, called the *domain* of a. The pair $\mathbb{A} = (U, A)$ is called an *information system* (see, e.g., [164]). Any information system can be represented by a data table with rows labeled by objects and columns labeled by attributes[3]. Any pair (x, a), where $x \in U$ and $a \in A$ defines the table entry consisting of the value $a(x)$.

Any subset B of A determines a binary relation $\mathcal{IND}_B$ on U, called an *indiscernibility relation*, defined by

$$x \; \mathcal{IND}_B \; y \text{ if and only if } a(x) = a(y) \text{ for every } a \in B, \tag{2.1}$$

where $a(x)$ denotes the value of attribute a for object x.

Obviously, $\mathcal{IND}_B$ is an equivalence relation. The family of all equivalence classes of $\mathcal{IND}_B$, i.e., the partition determined by B, will be denoted by $U / \mathcal{IND}_B$, or simply U/B; an equivalence class of $\mathcal{IND}_B$, i.e., the block of the partition U/B, containing

[3] Note, that in statistics or machine learning such a data table is called a sample [77].

x will be denoted by $B(x)$ (other notation used: $[x]_B$ or more precisely $[x]_{\mathit{IND}_B}$). Thus in view of the data we are unable, in general, to observe individual objects but we are forced to reason only about the accessible granules of knowledge (see, e.g., [153, 169, 206]).

If $(x,y) \in \mathit{IND}_B$ we will say that x and y are *B-indiscernible*. Equivalence classes of the relation IND_B (or blocks of the partition U/B) are referred to as *B-elementary sets* or *B-elementary granules*. In the rough set approach the elementary sets are the basic building blocks (concepts) of our knowledge about reality. The unions of *B-elementary sets* are called *B-definable sets*.[4]

For $B \subseteq A$ we denote by $Inf_B(x)$ the *B-signature* of $x \in U$, i.e., the set $\{(a,a(s)) : a \in B\}$. Let $INF(B) = \{Inf_B(s) : s \in U\}$. Then for any objects $x,y \in U$ the following equivalence holds: $x\mathit{IND}_B y$ if and only if $Inf_B(x) = Inf_B(y)$.

The indiscernibility relation will be further used to define basic concepts of rough set theory. Let us define now the following two operations on sets $X \subseteq U$

$$\mathsf{LOW}_B(X) = \{x \in U : B(x) \subseteq X\}, \tag{2.2}$$

$$\mathsf{UPP}_B(X) = \{x \in U : B(x) \cap X \neq \varnothing\}, \tag{2.3}$$

assigning to every subset X of the universe U two sets $\mathsf{LOW}_B(X)$ and $\mathsf{UPP}_B(X)$ called the *B-lower* and the *B-upper approximation* of X, respectively. The set

$$\mathsf{BN}_B(X) = \mathsf{UPP}_B(X) - \mathsf{LOW}_B(X), \tag{2.4}$$

will be referred to as the *B-boundary region* of X.

From the definition we obtain the following interpretation:

- The *lower approximation* of a set X with respect to B is the set of all objects, which can be for *certain* classified as objects in X using B (are *certainly* in X in view of B).
- The *upper approximation* of a set X with respect to B is the set of all objects which can be *possibly* classified as objects in X using B (are *possibly* in X in view of B).
- The *boundary region* of a set X with respect to B is the set of all objects, which can be classified neither as in X nor as in $U - X$ using B.

In other words, due to the granularity of knowledge, rough sets cannot be characterized by using available knowledge. Therefore with every rough set we associate two *crisp* sets, called *lower* and *upper approximation*. Intuitively, the lower approximation of a set consists of all elements that *surely* belong to the set, whereas the upper approximation of the set constitutes of all elements that *possibly* belong to the set, and the *boundary region* of the set consists of all elements that cannot be classified uniquely to the set or its complement, by employing available knowledge. The approximation definition is clearly depicted in Figure 2.1.

[4] One can compare data tables corresponding to information systems with relations in relational databases [52].

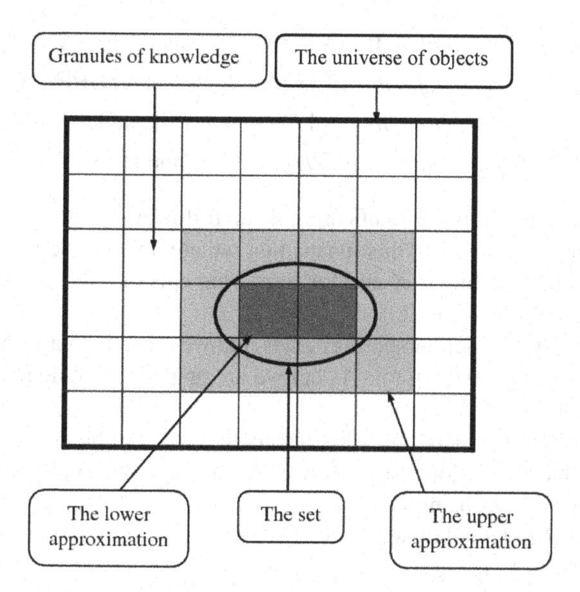

Fig. 2.1. A rough set.

The approximations have the following properties:

$$
\begin{aligned}
&\mathsf{LOW}_B(X) \subseteq X \subseteq \mathsf{UPP}_B(X), \\
&\mathsf{LOW}_B(\varnothing) = \mathsf{UPP}_B(\varnothing) = \varnothing, \mathsf{LOW}_B(U) = \mathsf{UPP}_B(U) = U, \\
&\mathsf{UPP}_B(X \cup Y) = \mathsf{UPP}_B(X) \cup \mathsf{UPP}_B(Y), \\
&\mathsf{LOW}_B(X \cap Y) = \mathsf{LOW}_B(X) \cap \mathsf{LOW}_B(Y), \\
&X \subseteq Y \text{ implies } \mathsf{LOW}_B(X) \subseteq \mathsf{LOW}_B(Y) \text{ and } \mathsf{UPP}_B(X) \subseteq \mathsf{UPP}_B(Y), \\
&\mathsf{LOW}_B(X \cup Y) \supseteq \mathsf{LOW}_B(X) \cup \mathsf{LOW}_B(Y), \\
&\mathsf{UPP}_B(X \cap Y) \subseteq \mathsf{UPP}_B(X) \cap \mathsf{UPP}_B(Y), \\
&\mathsf{LOW}_B(-X) = -\mathsf{UPP}_B(X), \\
&\mathsf{UPP}_B(-X) = -\mathsf{LOW}_B(X), \\
&\mathsf{LOW}_B(\mathsf{LOW}_B(X)) = \mathsf{UPP}_B(\mathsf{LOW}_B(X)) = \mathsf{LOW}_B(X), \\
&\mathsf{UPP}_B(\mathsf{UPP}_B(X)) = \mathsf{LOW}_B(\mathsf{UPP}_B(X)) = \mathsf{UPP}_B(X).
\end{aligned} \tag{2.5}
$$

Let us note that the inclusions in (2.5) cannot be in general substituted by the equalities. This has some important algorithmic and logical consequences.

Now we are ready to give the definition of rough sets.

If the boundary region of X is the empty set, i.e., $\mathsf{BN}_B(X) = \varnothing$, then the set X is *crisp* (*exact*) with respect to B; in the opposite case, i.e., if $\mathsf{BN}_B(X) \neq \varnothing$, the set X is referred to as *rough* (*inexact*) with respect to B. Thus any rough set, in contrast to a crisp set, has a non-empty boundary region.

One can define the following four basic classes of rough sets, i.e., four categories of vagueness:

$$X \text{ is } \textit{roughly B-definable} \text{ iff } \mathsf{LOW}_B(X) \neq \varnothing \text{ and } \mathsf{UPP}_B(X) \neq U, \tag{2.6}$$
$$X \text{ is } \textit{internally B-indefinable} \text{ iff } \mathsf{LOW}_B(X) = \varnothing \text{ and } \mathsf{UPP}_B(X) \neq U,$$
$$X \text{ is } \textit{externally B-indefinable} \text{ iff } \mathsf{LOW}_B(X) \neq \varnothing \text{ and } \mathsf{UPP}_B(X) = U,$$
$$X \text{ is } \textit{totally B-indefinable} \text{ iff } \mathsf{LOW}_B(X) = \varnothing \text{ and } \mathsf{UPP}_B(X) = U.$$

The intuitive meaning of this classification is the following.

If X is roughly *B-definable*, this means that we are able to decide for some elements of U that they belong to X and for some elements of U we are able to decide that they belong to $-X$, using B.

If X is internally B-indefinable, this means that we are able to decide about some elements of U that they belong to $-X$, but we are unable to decide for any element of U that it belongs to X, using B.

If X is externally B-indefinable, this means that we are able to decide for some elements of U that they belong to X, but we are unable to decide, for any element of U that it belongs to $-X$, using B.

If X is totally B-indefinable, we are unable to decide for any element of U whether it belongs to X or $-X$, using B.

Thus a set is *rough* (imprecise) if it has nonempty boundary region; otherwise the set is crisp (precise). This is exactly the idea of vagueness proposed by Frege.

Let us observe that the definition of rough sets refers to data (knowledge), and is *subjective*, in contrast to the definition of classical sets, which is in some sense an *objective* one.

A rough set can also be characterized numerically by the following coefficient

$$\alpha_B(X) = \frac{|\mathsf{LOW}_B(X)|}{|\mathsf{UPP}_B(X)|}, \tag{2.7}$$

called the *accuracy of approximation*, where X is a nonempty set and $|S|$ denotes the cardinality of set S.[5] Obviously $0 \leq \alpha_B(X) \leq 1$. If $\alpha_B(X) = 1$ then X is *crisp* with respect to B (X is *precise* with respect to B), and otherwise, if $\alpha_B(X) < 1$ then X is *rough* with respect to B (X is *vague* with respect to B). The accuracy of approximation can be used to measure the quality of approximation of decision classes on the universe U. One can use another measure of accuracy defined by $1 - \alpha_B(X)$ or by $1 - \frac{|\mathsf{BN}_B(X)|}{|U|}$. Some other measures of approximation accuracy are also used, e.g., based on entropy or some more specific properties of boundary regions (see, e.g., [53, 233, 259]). The choice of a relevant accuracy of approximation depends on a particular data set. Observe that the accuracy of approximation of X can be tuned by B. Another approach to accuracy of approximation can be based on the Variable Precision Rough Set Model (VPRSM) [321].

In the next section, we discuss decision rules (constructed over a selected set B of features or a family of sets of features) which are used in inducing classification algorithms (classifiers) making it possible to classify to decision classes unseen objects. Parameters which are tuned in searching for a classifier with the high quality

[5] The cardinality of set S is also denoted by "card(S)" instead of $|S|$.

are its description size (defined using decision rules) and its quality of classification (measured by the number of misclassified objects on a given set of objects). By selecting a proper balance between the accuracy of classification and the description size we expect to find the classifier with the high quality of classification also on unseen objects. This approach is based on the minimum description length principle [217, 218, 260].

2.3.2 Decision Systems and Decision Rules

Sometimes we distinguish in an information system $\mathbb{A} = (U,A)$ a partition of A into two disjoint classes $C,D \subseteq A$ of attributes, called *condition* and *decision (action)* attributes, respectively. The tuple $\mathbb{A} = (U,C,D)$ is called a *decision system* (or *decison table*[6]).

Let $V = \bigcup\{V_a \mid a \in C\} \bigcup\{V_d \mid d \in D\}$. Atomic formulae over $B \subseteq C \cup D$ and V are expressions $a = v$ called *descriptors (selectors) over B and V*, where $a \in B$ and $v \in V_a$. The set of formulae over B and V, denoted by $\mathscr{F}(B,V)$, is the least set containing all atomic formulae over B and V and closed under the propositional connectives $\wedge$ (conjunction), $\vee$ (disjunction) and $\neg$ (negation).

By $\|\varphi\|_{\mathbb{A}}$ we denote the meaning of $\varphi \in \mathscr{F}(B,V)$ in the decision system $\mathbb{A}$ which is the set of all objects in U with the property φ. These sets are defined by $\|a = v\|_{\mathbb{A}} = \{x \in U \mid a(x) = v\}$, $\|\varphi \wedge \varphi'\|_{\mathbb{A}} = \|\varphi\|_{\mathbb{A}} \cap \|\varphi'\|_{\mathbb{A}}$; $\|\varphi \vee \varphi'\|_{\mathbb{A}} = \|\varphi\|_{\mathbb{A}} \cup \|\varphi'\|_{\mathbb{A}}$; $\|\neg\varphi\|_{\mathbb{A}} = U - \|\varphi\|_{\mathbb{A}}$. The formulae from $\mathscr{F}(C,V)$, $\mathscr{F}(D,V)$ are called *condition formulae of* $\mathbb{A}$ and *decision formulae of* $\mathbb{A}$, respectively.

Any object $x \in U$ belongs to the *decision class* $\|\bigwedge_{d\in D} d = d(x)\|_{\mathbb{A}}$ of $\mathbb{A}$. All decision classes of $\mathbb{A}$ create a partition U/D of the universe U.

A *decision rule* for $\mathbb{A}$ is any expression of the form $\varphi \Rightarrow \psi$, where $\varphi \in \mathscr{F}(C,V)$, $\psi \in \mathscr{F}(D,V)$, and $\|\varphi\|_{\mathbb{A}} \neq \varnothing$. Formulae φ and ψ are referred to as the *predecessor* and the *successor* of decision rule $\varphi \Rightarrow \psi$. Decision rules are often called "*IF ...THEN ...*" rules. Such rules are used in machine learning (see, e.g., [77]).

Decision rule $\varphi \Rightarrow \psi$ is *true* in $\mathbb{A}$ if and only if $\|\varphi\|_{\mathbb{A}} \subseteq \|\psi\|_{\mathbb{A}}$. Otherwise, one can measure its *truth degree* by introducing some inclusion measure of $\|\varphi\|_{\mathbb{A}}$ in $\|\psi\|_{\mathbb{A}}$. Let us denote by $|\varphi|$ the number of objects from U that satisfies formula φ, i.e. the cardinality of $\|\varphi\|_{\mathbb{A}}$. According to Łukasiewicz [107], one can assign to formula φ the value $\frac{|\varphi|}{|U|}$, and to the implication $\varphi \Rightarrow \psi$ the fractional value $\frac{|\varphi \wedge \psi|}{|\varphi|}$, under the assumption that $\|\varphi\| \neq \varnothing$. Proposed by Łukasiewicz, that fractional part was much later adapted by machine learning and data mining literature, e.g. in the definitions of the accuracy of decision rules or confidence of association rules.

Each object x of a decision system determines a *decision rule*

$$\bigwedge_{a\in C} a = a(x) \Rightarrow \bigwedge_{d\in D} d = d(x). \tag{2.8}$$

For any decision system $\mathbb{A} = (U,C,D)$ one can consider a *generalized decision function* $\partial_A : U \longrightarrow \mathcal{P}(INF(D))$ defined by

[6] More precisely, decision tables are representations of decision systems.

$$\partial_{\mathbb{A}}(x) = \left\{ i \in INF(D) \; : \; \exists x' \in U \left[(x',x) \in IND_C \text{ and } Inf_D(x') = i \right] \right\}, \tag{2.9}$$

where $A = C \cup D$, $\mathcal{P}(INF(D))$ is the powerset of the set $INF(D)$ of all possible decision signatures.

The decision system $\mathbb{A}$ is called *consistent (deterministic)*, if $|\partial_{\mathbb{A}}(x)| = 1$, for any $x \in U$. Otherwise $\mathbb{A}$ is said to be *inconsistent (non-deterministic)*. Hence, decision system is inconsistent if it consists of some objects with different decisions but indiscernible with respect to condition attributes. Any set consisting of all objects with the same generalized decision value is called a *generalized decision class*.

Now, one can consider certain (possible) rules (see, e.g. [70, 72]) for decision classes defined by the lower (upper) approximations of such generalized decision classes of $\mathbb{A}$. This approach can be extend, using the relationships of rough sets with the Dempster-Shafer theory (see, e.g., [225, 233]), by considering rules relative to decision classes defined by the lower approximations of unions of decision classes of $\mathbb{A}$.

Numerous methods have been developed for generation of different types of decision rules, and the reader can find by himself in the literature on rough sets. Usually, one is searching for decision rules (semi) optimal with respect to some optimization criteria describing quality of decision rules in concept approximations.

In the case of searching for concept approximation in an extension of a given universe of objects (sample), the following steps are typical. When a set of rules has been induced from a decision system containing a set of training examples, they can be inspected to see if they reveal any novel relationships between attributes that are worth pursuing for further research. Furthermore, the rules can be applied to a set of unseen cases in order to estimate their classification power. For a systematic overview of rule application methods the reader is referred to the literature (see, e.g., [13, 115]).

2.3.3 Dependency of Attributes

Another important issue in data analysis is discovering dependencies between attributes in a given decision system $\mathbb{A} = (U,C,D)$. Intuitively, a set of attributes D depends totally on a set of attributes C, denoted $C \Rightarrow D$, if the values of attributes from C uniquely determine the values of attributes from D. In other words, D depends totally on C, if there exists a functional dependency between values of C and D. Hence, $C \Rightarrow D$ if and only if the rule (2.8) is true on $\mathbb{A}$ for any $x \in U$. In general, D can depend partially on C. Formally such a partial dependency can be defined in the following way.

We will say that D *depends on* C to a *degree* k $(0 \le k \le 1)$, denoted $C \Rightarrow_k D$, if

$$k = \gamma(C,D) = \frac{|POS_C(D)|}{|U|}, \tag{2.10}$$

where

$$POS_C(D) = \bigcup_{X \in U/D} \mathsf{LOW}_C(X), \tag{2.11}$$

called a *positive region* of the partition U/D with respect to C, is the set of all elements of U that can be uniquely classified to blocks of the partition U/D, by means of C.

If $k = 1$ we say that D *depends totally* on C, and if $k < 1$, we say that D *depends partially* (to *degree* k) on C. If $k = 0$ then the *positive region* of the partition U/D with respect to C is empty.

The coefficient k expresses the ratio of all elements of the universe, which can be properly classified to blocks of the partition U/D, employing attributes C and will be called the *degree of the dependency.*

It can be easily seen that if D depends totally on C then $\mathcal{IND}_C \subseteq \mathcal{IND}_D$. It means that the partition generated by C is finer than the partition generated by D. Notice, that the concept of dependency discussed above corresponds to that considered in relational databases.

Summing up: D is *totally* (*partially*) dependent on C, if *all* (*some*) elements of the universe U can be uniquely classified to blocks of the partition U/D, employing C.

Observe, that (2.10) defines only one of possible measures of dependency between attributes (see, e.g., [258]). One also can compare the dependency discussed in this section with dependencies considered in databases [52].

2.3.4 Reduction of Attributes

We often face a question whether we can remove some data from a data-table preserving its basic properties, that is – whether a table contains some superfluous data.

Let us express this idea more precisely.

Let $C, D \subseteq A$, be sets of condition and decision attributes respectively. We will say that $C' \subseteq C$ is a *D-reduct* (reduct with *respect* to D) of C, if C' is a minimal subset of C such that

$$\gamma(C,D) = \gamma(C',D). \tag{2.12}$$

The intersection of all D-reducts is called a *D-core* (core with *respect* to D). Because the core is the intersection of all reducts, it is included in every reduct, i.e., each element of the core belongs to some reduct. Thus, in a sense, the core is the most important subset of attributes, since none of its elements can be removed without affecting the classification power of attributes. Certainly, the geometry of reducts can be more compound. For example, the core can be empty but there can exist a partition of reducts into a few sets with non empty intersection.

Many other kinds of reducts and their approximations are discussed in the literature (see, e.g., [15, 130, 132, 226, 257, 259, 260, 94, 128, 274]). For example, if one change the condition (2.12) to $\partial_A(x) = \partial_B(x)$, (where $A = C \cup D$ and $B = C' \cup D$) then the defined reducts are preserving the generalized decision. Other kinds of reducts are preserving, e.g.: (i) the distance between attribute value vectors for any two objects, if this distance is greater than a given threshold [226], (ii) the distance between entropy distributions between any two objects, if this distance exceeds a given threshold [257, 259], or (iii) the so called reducts relative to object used for generation of decision rules [15]. There are some relationships between different kinds of reducts. If B is a reduct preserving the generalized decision, than in B is included a reduct preserving the positive region. For mentioned above reducts based on distances and thresholds one can find analogous dependency between reducts relative to different thresholds. By choosing different kinds of reducts we select different degrees to which information encoded in data is preserved. Reducts are used for building data models. Choosing a particular reduct or a set of reducts has impact on the model size as well as on its quality in describing a given data set. The model size together with the model quality are two basic components tuned in selecting relevant data models. This is known as the minimum length principle (see, e.g., [217, 218, 259, 260]). Selection of relevant kinds of reducts is an important step in building data models. It turns out that the different kinds of reducts can be efficiently computed using heuristics based, e.g., on the Boolean reasoning approach [25, 26, 24, 30].

2.3.5 Discernibility and Boolean Reasoning

Methodologies devoted to data mining, knowledge discovery, decision support, pattern classification, approximate reasoning require tools for discovering *templates (patterns)* in data and classifying them into certain *decision classes*. Templates are in many cases most frequent sequences of events, most probable events, regular configurations of objects, the decision rules of high quality, standard reasoning schemes. Tools for discovering and classifying of templates are based on *reasoning schemes* rooted in various paradigms [40]. Such patterns can be extracted from data by means of methods based, e.g., on Boolean reasoning and discernibility.

The discernibility relations are closely related to indiscernibility and belong to the most important relations considered in rough set theory.

The ability to discern between perceived objects is important for constructing many entities like reducts, decision rules or decision algorithms. In the standard approach the discernibility relation $\mathcal{DIS}_B \subseteq U \times U$ is defined by $x\ \mathcal{DIS}_B\ y$ if and only if $non(x\ \mathcal{IND}_B\ y)$, i.e., $B(x) \cap B(y) = \varnothing$. However, this is, in general, not the case for generalized approximation spaces. For example, in the case of some of such spaces, for any object x may be given a family $F(x)$ with more then one elementary granules (neighborhoods) such that $x \in \mathfrak{I}(x)$ for any $\mathfrak{I}(x) \in F(x)$. Then, one can define that objects x, y are discernible if and only if $\mathfrak{I}(x) \cap \mathfrak{I}(y) = \varnothing$ for some $\mathfrak{I}(x) \in F(x)$ and $\mathfrak{I}(y) \in F(y)$ and indiscernibility my be not the negation of this condition, e.g., objects x, y are defined as indiscernible if and only if $\mathfrak{I}(x) \cap \mathfrak{I}(y) \neq \varnothing$ for some $\mathfrak{I}(x) \in F(x)$ and $\mathfrak{I}(y) \in F(y)$.

The idea of Boolean reasoning is based on construction for a given problem P of a corresponding Boolean function f_P with the following property: the solutions for the problem P can be decoded from prime implicants of the Boolean function f_P. Let us mention that to solve real-life problems it is necessary to deal with Boolean functions having large number of variables.

A successful methodology based on the discernibility of objects and Boolean reasoning has been developed for computing of many important ingredients for applications. These applications include generation of reducts and their approximations, decision rules, association rules, discretization of real value attributes, symbolic value grouping, searching for new features defined by oblique hyperplanes or higher order surfaces, pattern extraction from data as well as conflict resolution or negotiation (see, e.g., [15, 130, 132, 226, 257, 259, 260, 128]).

Most of the problems related to generation of the above mentioned entities are NP-complete or NP-hard. However, it was possible to develop efficient heuristics returning suboptimal solutions of the problems. The results of experiments on many data sets are very promising. They show very good quality of solutions generated by the heuristics in comparison with other methods reported in literature (e.g., with respect to the classification quality of unseen objects). Moreover, they are very efficient from the point of view of time necessary for computing of the solution. Many of these methods are based on discernibility matrices. Note, that it is possible to compute the necessary information about these matrices using directly[7] information or decision systems (e.g., sorted in preprocessing [13, 127, 134, 307]) which significantly improves the efficiency of algorithms.

It is important to note that the methodology makes it possible to construct heuristics having a very important *approximation property* which can be formulated as follows: expressions, called *approximate implicants*, generated by heuristics that are *close* to prime implicants define approximate solutions for the problem.

2.3.6 Rough Membership

Let us observe that rough sets can be also defined employing the rough membership function (see Eq. 2.13) instead of approximation [172]. That is, consider

$$\mu_X^B : U \to [0,1],$$

defined by

$$\mu_X^B(x) = \frac{|B(x) \cap X|}{|X|}, \tag{2.13}$$

where $x \in X \subseteq U$. The value $\mu_X^B(x)$ can be interpreted as the degree that x belongs to X in view of knowledge about x expressed by B or the degree to which the elementary granule $B(x)$ is included in the set X. This means that the definition reflects a subjective knowledge about elements of the universe, in contrast to the classical definition of a set.

The rough membership function can also be interpreted as the conditional probability that x belongs to X given B. This interpretation was used by several researchers in the rough set community (see, e.g., [71, 259, 293, 306, 325, 314], [321]). Note also that the ratio on the right hand side of the equation (2.13) is known as the confidence coefficient in data mining [77, 89]. It is worthwhile to mention that set inclusion to a degree has been considered by Łukasiewicz [107] in studies on assigning fractional truth values to logical formulas.

On can observe that the rough membership function has the following properties [172]:

[7] i.e., without the necessity of generation and storing of the discernibility matrices.

1) $\mu_X^B(x) = 1$ *iff* $x \in \mathsf{LOW}_B(X)$,
2) $\mu_X^B(x) = 0$ *iff* $x \in U - \mathsf{UPP}_B(X)$,
3) $0 < \mu_X^B(x) < 1$ *iff* $x \in \mathsf{BN}_B(X)$,
4) $\mu_{U-X}^B(x) = 1 - \mu_X^B(x)$ *for any* $x \in U$,
5) $\mu_{X\cup Y}^B(x) \geq \max\left(\mu_X^B(x),\, \mu_Y^B(x)\right)$ *for any* $x \in U$,
6) $\mu_{X\cap Y}^B(x) \leq \min\left(\mu_X^B(x),\, \mu_Y^B(x)\right)$ *for any* $x \in U$.

From the properties it follows that the rough membership differs essentially from the fuzzy membership [317], for properties 5) and 6) show that the membership for union and intersection of sets, in general, cannot be computed – as in the case of fuzzy sets – from their constituents membership. Thus formally the rough membership is different from fuzzy membership. Moreover, the rough membership function depends on an available knowledge (represented by attributes from B). Besides, the rough membership function, in contrast to fuzzy membership function, has a probabilistic flavor.

Let us also mention that rough set theory, in contrast to fuzzy set theory, clearly distinguishes two very important concepts, vagueness and uncertainty, very often confused in the AI literature. Vagueness is the property of concepts. Vague concepts can be approximated using the rough set approach [232]. Uncertainty is the property of elements of a set or a set itself (e.g., only examples and/or counterexamples of elements of a considered set are given). Uncertainty of elements of a set can be expressed by the rough membership function.

Both fuzzy and rough set theory represent two different approaches to vagueness. Fuzzy set theory addresses *gradualness* of knowledge, expressed by the fuzzy membership, whereas rough set theory addresses *granularity* of knowledge, expressed by the indiscernibility relation. A nice illustration of this difference has been given by Dider Dubois and Henri Prade [39] in the following example. In image processing fuzzy set theory refers to gradualness of gray level, whereas rough set theory is about the size of pixels.

Consequently, both theories are not competing but are rather complementary. In particular, the rough set approach provides tools for approximate construction of fuzzy membership functions. The rough-fuzzy hybridization approach proved to be successful in many applications (see, e.g., [150, 154]).

Interesting discussion of fuzzy and rough set theory in the approach to vagueness can be found in [216]. Let us also observe that fuzzy set and rough set theory are not a remedy for classical set theory difficulties.

One of the consequences of perceiving objects by information about them is that for some objects one cannot decide if they belong to a given set or not. However, one can estimate the degree to which objects belong to sets. This is a crucial observation in building foundations for approximate reasoning. Dealing with imperfect knowledge implies that one can only characterize satisfiability of relations between objects to a degree, not precisely. One of the fundamental relations on objects is a rough inclusion relation describing that objects are parts of other objects to a degree. The rough mereological approach [199, 153, 200, 201, 203] based on such a relation is an extension of the Leśniewski mereology [98].

2.4 Generalizations of Approximation Spaces

The rough set concept can be defined quite generally by means of topological operations, *interior* and *closure*, called *approximations* [196]. It was observed in [166] that the key to the presented approach is provided by the exact mathematical formulation of the concept of approximative (rough) equality of sets in a given approximation space. In [169], an approximation space is represented by the pair $(U, \mathcal{R})$, where U is a universe of objects, and $\mathcal{R} \subseteq U \times U$ is an indiscernibility relation defined by an attribute set (i.e., $\mathcal{R} = \mathcal{IND}_A$ for some attribute set A). In this case $\mathcal{R}$ is the equivalence relation. Let $[x]_{\mathcal{R}}$ denote an equivalence class of an element $x \in U$ under the indiscernibility relation $\mathcal{R}$, where $[x]_{\mathcal{R}} = \{y \in U : x\mathcal{R}y\}$.

In this context, $\mathcal{R}$-approximations of any set $X \subseteq U$ are based on the exact (crisp) containment of sets. Then set approximations are defined as follows:

- $x \in U$ belongs with certainty to $X \subseteq U$ (i.e., x belongs to the $\mathcal{R}$-lower approximation of X), if $[x]_{\mathcal{R}} \subseteq X$.
- $x \in U$ possibly belongs $X \subseteq U$ (i.e., x belongs to the $\mathcal{R}$-upper approximation of X), if $[x]_{\mathcal{R}} \cap X \neq \emptyset$.
- $x \in U$ belongs with certainty neither to the X nor to $U - X$ (i.e., x belongs to the $\mathcal{R}$-boundary region of X), if $[x]_{\mathcal{R}} \cap (U - X) \neq \emptyset$ and $[x]_{\mathcal{R}} \cap X \neq \emptyset$.

Our knowledge about the approximated concepts is often partial and uncertain [69]. For example, concept approximation should be constructed from examples and counterexamples of objects for the concepts [77]. Hence, concept approximations constructed from a given sample of objects are extended, using inductive reasoning, on objects not yet observed.

Several generalizations of the classical rough set approach based on approximation spaces defined as pairs of the form $(U, \mathcal{R})$, where $\mathcal{R}$ is the equivalence relation (called indiscernibility relation) on the set U, have been reported in the literature (see, e.g., [100, 102, 202, 229, 292, 311, 312, 313, 315, 251, 245, 272, 244, 246]) [8].

Let us mention two of them.

The concept approximations should be constructed under dynamically changing environments [232, 250]. This leads to a more complex situation where the boundary regions are not crisp sets, which is consistent with the postulate of the higher order vagueness considered by philosophers (see, e.g., [87]). Different aspects of vagueness in the rough set framework are discussed, e.g., in [109, 140, 142, 216, 232]. It is worthwhile to mention that a rough set approach to the approximation of compound concepts has been developed. For such concepts, it is hardly possible to expect that they can be approximated with the high quality by the traditional methods [29, 300]. The approach is based on hierarchical learning and ontology approximation [11, 19, 133, 153, 237]. Approximation of concepts in distributed environments is discussed in [230]. A survey of algorithmic methods for concept approximation based on rough sets and Boolean reasoning in presented, e.g., in [13, 227, 128, 11].

[8] Among extensions not discussed in this chapter is the rough set approach to multicriteria decision making (see, e.g., [61, 62, 63, 64, 68, 183, 194, 267, 65, 93, 66, 67, 60, 265]).

A generalized approximation space[9] can be defined by a tuple $\mathbb{AS} = (U, \mathcal{I}, \nu)$ where $\mathcal{I}$ is the *uncertainty function* defined on U with values in the powerset $\mathcal{P}(U)$ of U ($\mathcal{I}(x)$ is the *neighboorhood* of x) and ν is the *inclusion function* defined on the Cartesian product $\mathcal{P}(U) \times \mathcal{P}(U)$ with values in the interval $[0,1]$ measuring the degree of inclusion of sets [240]. The lower and upper approximation operations can be defined in $\mathbb{AS}$ by

$$\mathsf{LOW}_{\mathbb{AS}}(X) = \{x \in U : \nu(\mathcal{I}(x), X) = 1\}, \tag{2.14}$$

$$\mathsf{UPP}_{\mathbb{AS}}(X) = \{x \in U : \nu(\mathcal{I}(x), X) > 0\}. \tag{2.15}$$

In the standard case, $\mathcal{I}(x)$ is equal to the equivalence class $B(x)$ of the indiscernibility relation IND_B; in case of tolerance (similarity) relation $\mathcal{T} \subseteq U \times U$ [207] we take $\mathcal{I}(x) = [x]_{\mathcal{T}} = \{y \in U : x\ \mathcal{T}\ y\}$, i.e., $\mathcal{I}(x)$ is equal to the tolerance class of $\mathcal{T}$ defined by x. The standard rough inclusion relation ν_{SRI} is defined for $X, Y \subseteq U$ by

$$\nu_{SRI}(X,Y) = \begin{cases} \frac{|X \cap Y|}{|X|}, & \text{if } X \text{ is non} - \text{empty}, \\ 1, & \text{otherwise.} \end{cases} \tag{2.16}$$

For applications it is important to have some constructive definitions of $\mathcal{I}$ and ν.

One can consider another way to define $\mathcal{I}(x)$. Usually together with $\mathbb{AS}$ we consider some set $\mathscr{F}$ of formulae describing sets of objects in the universe U of $\mathbb{AS}$ defined by semantics $\|\cdot\|_{\mathbb{AS}}$, i.e., $\|\alpha\|_{\mathbb{AS}} \subseteq U$ for any $\alpha \in \mathscr{F}$[10]. Now, one can take the set

$$N_{\mathscr{F}}(x) = \{\alpha \in \mathscr{F} : x \in \|\alpha\|_{\mathbb{AS}}\}, \tag{2.17}$$

and $\mathcal{I}_\circ(x) = \{\|\alpha\|_{\mathbb{AS}} : \alpha \in N_{\mathscr{F}}(x)\}$. Hence, more general uncertainty functions having values in $\mathcal{P}(\mathcal{P}(U))$ can be defined and in the consequence different definitions of approximations are considered. For example, one can consider the following definitions of approximation operations in this approximation space $\mathbb{AS}_\circ = (U, \mathcal{I}_\circ, \nu)$:

$$\mathsf{LOW}_{\mathbb{AS}_\circ}(X) = \{x \in U : \nu(Y, X) = 1 \text{ for some } Y \in \mathcal{I}(x)\}, \tag{2.18}$$

$$\mathsf{UPP}_{\mathbb{AS}_\circ}(X) = \{x \in U : \nu(Y, X) > 0 \text{ for any } Y \in \mathcal{I}(x)\}. \tag{2.19}$$

There are also different forms of rough inclusion functions. Let us consider two examples.

In the first example of a rough inclusion function, a threshold $t \in (0, 0.5)$ is used to relax the degree of inclusion of sets. The rough inclusion function ν_t is defined by

$$\nu_t(X,Y) = \begin{cases} 1 & \text{if } \nu_{SRI}(X,Y) \geq 1-t, \\ \frac{\nu_{SRI}(X,Y)-t}{1-2t} & \text{if } t \leq \nu_{SRI}(X,Y) < 1-t, \\ 0 & \text{if } \nu_{SRI}(X,Y) \leq t. \end{cases} \tag{2.20}$$

---

[9] More general cases are considered, e.g., in [244, 246].

[10] If $\mathbb{AS} = (U, \mathcal{I}, \nu)$ then we will also write $\|\alpha\|_U$ instead of $\|\alpha\|_{\mathbb{AS}}$.

This is an interesting "rough-fuzzy" example because we put the standard rough membership function as an argument into the formula often used for fuzzy membership functions.

One can obtain approximations considered in the variable precision rough set approach (VPRSM) [321] by substituting in (2.14)-(2.15) the rough inclusion function ν_t defined by (2.20) instead of ν, assuming that Y is a decision class and $\mathfrak{I}(x) = B(x)$ for any object x, where B is a given set of attributes.

Another example of application of the standard inclusion was developed by using probabilistic decision functions. For more detail the reader is referred to [246, 285, 258, 259].

The rough inclusion relation can be also used for function approximation [251, 244, 246] and relation approximation [271]. In the case of function approximation the inclusion function ν^* for subsets $X,Y \subseteq U \times U$, where $U \subseteq \mathbb{R}$ and $\mathbb{R}$ is the set of real numbers, is defined by

$$\nu^*(X,Y) = \begin{cases} \frac{|\pi_1(X \cap Y)|}{|\pi_1(X)|} & \text{if } \pi_1(X) \neq \varnothing, \\ 1 & \text{if } \pi_1(X) = \varnothing, \end{cases} \quad (2.21)$$

where π_1 is the projection operation on the first coordinate. Assume now, that X is a cube and Y is the graph $G(f)$ of the function $f : \mathbb{R} \longrightarrow \mathbb{R}$. Then, e.g., X is in the lower approximation of f if the projection on the first coordinate of the intersection $X \cap G(f)$ is equal to the projection of X on the first coordinate. This means that the part of the graph $G(f)$ is "well" included in the box X, i.e., for all arguments that belong to the box projection on the first coordinate the value of f is included in the box X projection on the second coordinate. This approach was extended in several papers (see, e.g., [285, 246]).

The approach based on inclusion functions has been generalized to the *rough mereological approach* [153, 201, 200, 203]. The inclusion relation $x\mu_r y$ with the intended meaning *x is a part of y to a degree at least r* has been taken as the basic notion of the rough mereology being a generalization of the Leśniewski mereology [98, 99]. Research on rough mereology has shown importance of another notion, namely *closeness* of compound objects (e.g., concepts). This can be defined by $x \; cl_{r,r'} \; y$ if and only if $x \; \mu_r \; y$ and $y \; \mu_{r'} \; x$.

Rough mereology offers a methodology for synthesis and analysis of objects in a distributed environment of intelligent agents, in particular, for synthesis of objects satisfying a given specification to a satisfactory degree or for control in such a complex environment. Moreover, rough mereology has been used for developing the foundations of the *information granule calculi*, aiming at formalization of the Computing with Words paradigm, formulated by Lotfi Zadeh [318]. More complex information granules are defined recursively using already defined information granules and their measures of inclusion and closeness. Information granules can have complex structures like classifiers or approximation spaces. Computations on information granules are performed to discover relevant information granules, e.g., patterns or approximation spaces for compound concept approximations.

Usually there are considered families of approximation spaces labeled by some parameters. By tuning such parameters according to chosen criteria (e.g., minimal description length) one can search for the optimal approximation space for concept description (see, e.g., [13, 128, 11]).

2.5 Rough Sets and Induction

Granular formulas are constructed from atomic formulas corresponding to the considered attributes (see, e.g., [175, 174, 244, 246]). In the consequence, the satisfiability of such formulas is defined if the satisfiability of atomic formulas is given as the result of sensor measurement. Let us consider two information systems $\mathbb{A} = (U, C, D)$ and its extension $\mathbb{A}^* = (U^*, C)$ having the same set of attributes C (more precisely, the set o attributes in $\mathbb{A}$ is obtained by restricting to U attributes from $\mathbb{A}^*$ defined on $U^* \supseteq U$). Hence, one can consider for any constructed formula α over atomic formulas its semantics $\|\alpha\|_{\mathbb{A}} \subseteq U$ over U as well as the semantics $\|\alpha\|_{\mathbb{A}^*} \subseteq U^*$ over U^* (see Figure 2.2).

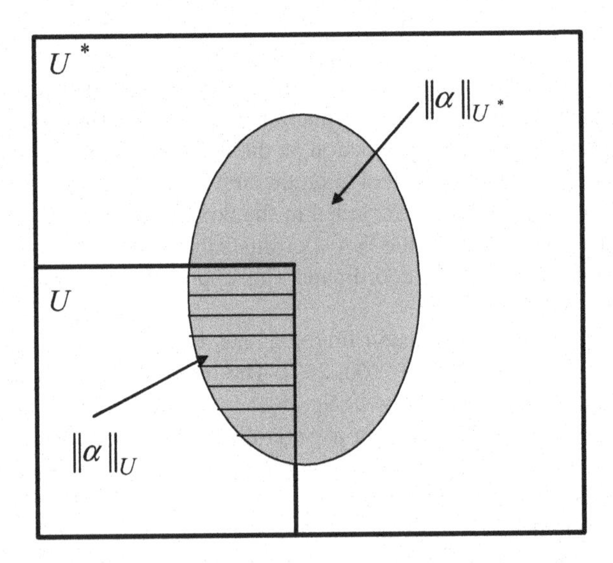

Fig. 2.2. Two semantics of α over U and U^*, respectively.

The difference between these two cases is the following. In the case of U, one can compute $\|\alpha\|_{\mathbb{A}} \subseteq U$ but in the case $\|\alpha\|_{\mathbb{A}^*} \subseteq U^*$, for any object from $U^* - U$, there is no information about its membership relative to $\|\alpha\|_{\mathbb{A}^*} - \|\alpha\|_{\mathbb{A}}$. One can estimate the satisfiability of α for objects $u \in U^* - U$ only after some relevant sensory measurements on u are performed. In particular, one can use some methods for estimation of relationships among semantics of formulas over U^* using the relationships among semantics of these formulas over U. For example, one can apply statistical methods. This step is crucial in investigation of extensions of approximation spaces relevant for inducing classifiers from data.

2.5.1 Rough Sets and Classifiers

In this section, we consider the problem of approximation of concepts over a universe U^∞ (concepts that are subsets of U^∞). We assume that the concepts are perceived only through some subsets of U^∞, called samples. This is a typical situation in the machine learning, pattern recognition, or data mining approaches [77, 89, 103]. We explain the rough set approach to induction of concept approximations using the generalized approximation spaces of the form $\mathbb{AS} = (U, I, \nu)$ defined in Section 2.4.

Let $U \subseteq U^\infty$ be a finite sample. By Π_U we denote a perception function from $\mathcal{P}(U^\infty)$ into $\mathcal{P}(U)$ defined by $\Pi_U(C) = C \cap U$ for any concept $C \subseteq U^\infty$.

Let us consider first an illustrative example.

We assume that there is given an information system $\mathbb{A} = (U, A)$ and let us assume that for some $C \subseteq U^\infty$ there is given the set $\Pi_U(C) = C \cap U$. In this way we obtain a decision system $\mathbb{AS}_d = (U, A, d)$, where $d(x) = 1$ if $x \in C \cap U$ and $d(x) = 0$, otherwise.

We would like to illustrate how from the decision function d may be induced a decision function d^* defined over U^∞ which can be treated as an approximation of the characteristic function of C.

Let us assume that $RULES(\mathbb{AS}_d)$ is a set of decision rules induced by some rule generation method from $\mathbb{AS}_d$. For any object $x \in U^\infty$, let $MatchRules(\mathbb{AS}_d, x)$ be the set of rules from $RULES(\mathbb{AS}_d)$ supported by x (see, e.g., [15]).

Let $C_1 = C$ and $C_0 = U^\infty \setminus C$. Now, for $k = 1, 0$ one can define the rough membership functions $\mu_k : U^\infty \to [0, 1]$ in the following way:

1. Let $R_k(x)$ be the set of all decision rules from $MatchRules(\mathbb{AS}_d, x)$ for C_k, i.e., decision rules from $MatchRules(\mathbb{AS}_d, x)$ with right hand side $d = k$
2. We define real values $w_k(x)$, where $w_1(x)$ is called the weight "for" and $w_0(x)$ the weight "against" membership of the object $x \in C_k$, respectively, by

$$w_k(x) = \sum_{r \in R_k(x)} strength(r),$$

   where $strength(r)$ is a normalized function depending on $length$, $support$, $confidence$ of the decision rule r and on some global information about the decision table $\mathbb{AS}_d$ such as the table size or the class distribution (see [15]).
3. Finally, one can define the value of $\mu_k(x)$ by

$$\mu_k(x) = \begin{cases} \text{undefined} & \text{if } \max(w_k(x), w_{1-k}(x)) < \omega \\ 0 & \text{if } w_{1-k}(x) - w_k(x) \geq \theta \text{ and } w_{1-k}(x) > \omega \\ 1 & \text{if } w_k(x) - w_{1-k}(x) \geq \theta \text{ and } w_k(x) > \omega \\ \frac{\theta + (w_k(x) - w_{1-k}(x))}{2\theta} & \text{in other cases,} \end{cases}$$

   where ω, θ are parameters set by user.

Now, for computing of the value $d^*(x)$ for $x \in U^\infty$ the user should select a strategy resolving conflicts between values $\mu_1(x)$ and $\mu_0(x)$ representing, in a sense votes "for" and "against" membership of x in C, respectively. Note that for some cases

x due to the small differences between these values the selected strategy may not produce the definite answer and these cases will create the boundary region.

Let us consider a generalized approximation space $\mathbb{AS} = (U, I, \nu_{SRI})$, where $I(x) = A(x)$ for $x \in U$. Now we would like to look how this approximation space may be inductively extended so that in the induced approximation space we may define approximation of the concept C or in other words the approximation of the decision function d^*.

Hence, the problem we are considering is how to extend the approximations of $\Pi_U(C)$ defined by $\mathbb{AS}$ to an approximation of C over U^∞. We show that the problem can be described as searching for an extension $\mathbb{AS}_C = (U^\infty, \mathfrak{I}_C, \nu_C)$ of the approximation space $\mathbb{AS}$, relevant for approximation of C. This requires to show how to extend the inclusion function ν from subsets of U to subsets of U^∞ that are relevant for the approximation of C. Observe that for the approximation of C it is enough to induce the necessary values of the inclusion function ν_C without knowing the exact value of $\mathfrak{I}_C(x) \subseteq U^\infty$ for $x \in U^\infty$.

Let $\mathbb{AS}$ be a given approximation space for $\Pi_U(C)$ and let us consider a language L in which the neighborhood $\mathfrak{I}(x) \subseteq U$ is expressible by a formula $pat(x)$, for any $x \in U$. It means that $\mathfrak{I}(x) = \|pat(x)\|_U \subseteq U$, where $\|pat(x)\|_U$ denotes the meaning of $pat(x)$ restricted to the sample U. In case of rule based classifiers patterns of the form $pat(x)$ are defined by feature value vectors.

We assume that for any new object $x \in U^\infty - U$ we can obtain (e.g., as a result of sensor measurement) a pattern $pat(x) \in L$ with semantics $\|pat(x)\|_{U^\infty} \subseteq U^\infty$. However, the relationships between information granules over U^∞ like sets: $\|pat(x)\|_{U^\infty}$ and $\|pat(y)\|_{U^\infty}$, for different $x, y \in U^\infty$, are, in general, known only if they can be expressed by relationships between the restrictions of these sets to the sample U, i.e., between sets $\Pi_U(\|pat(x)\|_{U^\infty})$ and $\Pi_U(\|pat(y)\|_{U^\infty})$.

The set of patterns $\{pat(x) : x \in U\}$ is usually not relevant for approximation of the concept $C \subseteq U^\infty$. Such patterns are too specific or not enough general, and can directly be applied only to a very limited number of new objects. However, by using some generalization strategies, one can search, in a family of patterns definable from $\{pat(x) : x \in U\}$ in L, for such new patterns that are relevant for approximation of concepts over U^∞. Let us consider a subset $PATTERNS(\mathbb{AS}, L, C) \subseteq L$ chosen as a set of pattern candidates for relevant approximation of a given concept C. For example, in case of rule based classifier one can search for such candidate patterns among sets definable by subsequences of feature value vectors corresponding to objects from the sample U. The set $PATTERNS(\mathbb{AS}, L, C)$ can be selected by using some quality measures checked on meanings (semantics) of its elements restricted to the sample U (like the number of examples from the concept $\Pi_U(C)$ and its complement that support a given pattern). Then, on the basis of properties of sets definable by these patterns over U we induce approximate values of the inclusion function ν_C on subsets of U^∞ definable by any of such pattern and the concept C.

Next, we induce the value of ν_C on pairs (X, Y) where $X \subseteq U^\infty$ is definable by a pattern from $\{pat(x) : x \in U^\infty\}$ and $Y \subseteq U^\infty$ is definable by a pattern from $PATTERNS(\mathbb{AS}, L, C)$.

Finally, for any object $x \in U^\infty - U$ we induce the approximation of the degree

$$\nu_C(\|pat(x)\|_{U^\infty}, C)$$

applying a conflict resolution strategy $Conflict\_res$ (a voting strategy, in case of rule based classifiers) to two families of degrees:

$$\{\nu_C(\|pat(x)\|_{U^\infty}, \|pat\|_{U^\infty}) : pat \in PATTERNS(\mathbb{AS}, L, C)\}, \quad (2.22)$$

$$\{\nu_C(\|pat\|_{U^\infty}, C) : pat \in PATTERNS(\mathbb{AS}, L, C)\}. \quad (2.23)$$

Values of the inclusion function for the remaining subsets of U^∞ can be chosen in any way – they do not have any impact on the approximations of C. Moreover, observe that for the approximation of C we do not need to know the exact values of uncertainty function I_C – it is enough to induce the values of the inclusion function ν_C. Observe that the defined extension ν_C of ν to some subsets of U^∞ makes it possible to define an approximation of the concept C in a new approximation space $\mathbb{AS}_C$.

Observe that one can also follow principles of Bayesian reasoning and use degrees of ν_C to approximate C (see, e.g., [171, 261, 264]).

Let us present yet another example of (inductive) extension $\mathbb{AS}^*$ of approximation space $\mathbb{AS}$ in the case of rule based classifiers. For details the reader is referred to, e.g., [244, 246].

Let $h : [0,1] \to \{0, 1/2, 1\}$ be a function defined by

$$h(t) = \begin{cases} 1, & \text{if } t > \frac{1}{2}, \\ \frac{1}{2}, & \text{if } t = \frac{1}{2}, \\ 0, & \text{if } t < \frac{1}{2}. \end{cases} \quad (2.24)$$

We start with an extension of the uncertainty function and the rough inclusion function from U to U^*, where $U \subseteq U^*$:

$$\mathcal{I}(x) = \{\|lh(r)\|_{U^*} : x \in \|lh(r)\|_{U^*} \text{ and } r \in Rule\_set\}, \quad (2.25)$$

where $x \in U^*$ and $lh(r)$ denotes the formula on the left hand side of the rule r, and $Rule\_set$ is a set of decision rules induced from a given decision system $DT = (U, A, d)$. In this approach, the rough inclusion function is defined by

$$\nu_U(X, Z) = h\left(\frac{|\{Y \in X : Y \cap U \subseteq Z\}|}{|\{Y \in X : Y \cap U \subseteq Z\}| + |\{Y \in X : Y \cap U \subseteq U^* - Z\}|}\right), \quad (2.26)$$

where $X \subseteq \mathcal{P}(U^*)$, $X \neq \varnothing$ and $Z \subseteq U^*$. In case $X = \varnothing$ we set $\nu_U(\varnothing, Z) = 0$

The induced uncertainty and rough inclusion functions can now be used to define the lower approximation $\mathsf{LOW}_{\mathbb{AS}^*}(Z)$, the upper approximation $\mathsf{UPP}_{\mathbb{AS}^*}(Z)$, and the boundary region $\mathsf{BN}_{\mathbb{AS}^*}(Z)$ of $Z \subseteq U^*$ by:

$$\mathsf{LOW}_{\mathbb{AS}^*}(Z) = \{x \in U^* : \nu_U(\mathcal{I}(x), Z) = 1\}, \quad (2.27)$$

$$\mathsf{UPP}_{\mathbb{AS}^*}(Z) = \{x \in U^* : \nu_U(\mathcal{I}(x), Z) > 0\}, \quad (2.28)$$

and

$$\mathsf{BN}_{\mathbb{AS}^*}(Z) = \mathsf{UPP}_{\mathbb{AS}^*}(Z) - \mathsf{LOW}_{\mathbb{AS}^*}(Z). \quad (2.29)$$

In the example, we classify objects from U^* to the lower approximation of Z if majority of rules matching this object are voting for Z and to the upper approximation of Z if at least half of the rules matching x are voting for Z. Certainly, one can follow many other voting schemes developed in machine learning or by introducing less crisp conditions in the boundary region definition. The defined approximations can be treated as estimations of the exact approximations of subsets of U^* because they are induced on the basis of samples of such sets restricted to U only. One can use some standard quality measures developed in machine learning to calculate the quality of such approximations assuming that after estimation of approximations on U^* full information about membership for objects relative to the approximated subsets of U^* is uncovered analogously to the testing sets in machine learning.

In an analogous way, one can describe other class of classifiers used in machine learning and data mining such as neural networks or k-nn classifiers.

In this way, the rough set approach to induction of concept approximations can be explained as a process of inducing a relevant approximation space.

In [246] is presented the rough set approach to approximation of partially defined concepts (see also, e.g., [17, 18, 21, 245, 133, 128, 11, 272, 254, 244, 255]). The problems discussed in this chapter are crucial for building computer systems that assist researchers in scientific discoveries in many areas. Our considerations can be treated as a step toward foundations for modeling of granular computations inside of system that is based on granules called approximation spaces. Approximation spaces are fundamental granules used in searching for relevant complex granules called as data models, e.g., approximations of complex concepts, functions or relations. The approach requires some generalizations of the approximation space concept introduced in [239, 240]. There are presented examples of rough set-based strategies for the extension of approximation spaces from samples of objects onto a whole universe of objects. This makes it possible to present foundations for inducing data models such as approximations of concepts or classifications analogous to the approaches for inducing different types of classifiers known in machine learning and data mining. Searching for relevant approximation spaces and data models are formulated as complex optimization problems. This optimization is performed relative to some measures which are some versions of the minimum length principle (MLP) [217, 218].

2.5.2 Inducing Relevant Approximation Spaces

A key task in granular computing is the information granulation process that leads to the formation of information aggregates (with inherent patterns) from a set of available objects. A methodological and algorithmic issue is the formation of transparent (understandable) information granules inasmuch as they should provide a clear and understandable description of patterns present in sample objects [7, 176]. Such a fundamental property can be formalized by a set of constraints that must be satisfied during the information granulation process. For example, in case of inducing granules such as classifiers, the constraints specify requirements for the quality of classifiers. Then, inducing of classifiers can be understood as searching for relevant approximation spaces (which can be treated as a spacial type of granules) relative to some properly selected optimization measures[11]. The selection of these optimization measures is not an easy task because they should guarantee that the (semi-) optimal approximation spaces selected relative to these criteria allow us to construct classifiers of the high quality.

Let us consider some examples of optimization measures [129]. For example, the quality of an approximation space can be measured by:

$$Quality_1 : \mathcal{SAS}(U) \times \mathcal{P}(U) \rightarrow [0,1], \tag{2.30}$$

where U is a non-empty set of objects and $\mathcal{SAS}(U)$ is a set of possible approximation spaces with the universe U.

Example 2.1. If $\mathsf{UPP}_{\mathbb{AS}}(X) \neq \varnothing$ for $\mathbb{AS} \in \mathcal{SAS}(U)$ and $X \subseteq U$ then

$$Quality_1(\mathbb{AS},X) = \nu_{SRI}(\mathsf{UPP}_{\mathbb{AS}}(X),\mathsf{LOW}_{\mathbb{AS}}(X)) = \frac{|\mathsf{LOW}_{\mathbb{AS}}(X)|}{|\mathsf{UPP}_{\mathbb{AS}}(X)|}. \tag{2.31}$$

The value $1 - Quality_1(\mathbb{AS},X)$ expresses the degree of completeness of our knowledge about X, given the approximation space $\mathbb{AS}$.

Example 2.2. In applications, we usually use another quality measure analogous to the minimum length principle [218, 218] where also the description length of approximation is included. Let us denote by $description(\mathbb{AS},X)$ the description length of approximation of X in $\mathbb{AS}$. The description length may be measured, e.g., by the sum of description lengths of algorithms testing membership for neighborhoods used in construction of the lower approximation, the upper approximation, and the boundary region of the set X. Then the quality $Quality_2(\mathbb{AS},X)$ can be defined by

$$Quality_2(\mathbb{AS},X) = g(Quality_1(\mathbb{AS},X), description(\mathbb{AS},X)), \tag{2.32}$$

where g is a relevant function used for fusion of values $Quality_1(\mathbb{AS},X)$ and $description(\mathbb{AS},X)$. This function g can reflect weights given by experts relative to both criteria.

One can consider different optimization problems relative to a given class $Set\_\mathbb{AS}$ of approximation spaces. For example, for a given $X \subseteq U$ and a threshold $t \in [0,1]$, one can search for an approximation space $\mathbb{AS}$ satisfying the constraint $Quality_2(\mathbb{AS},X) \geq t$.

Another example can be related to searching for an approximation space satisfying additionally the constraint $Cost(\mathbb{AS}) < c$ where $Cost(\mathbb{AS})$ denotes the cost of approximation space $\mathbb{AS}$ (*e.g.*, measured by the number of attributes used to define neighborhoods in $\mathbb{AS}$) and c is a given threshold. In the following example, we

[11] Note that while there is a large literature on the covering based rough set approach (see, e.g., [320, 73]) still much more work should be done on (scalable) algorithmic searching methods for relevant approximation spaces in huge families of approximation spaces defined by many parameters determining neighborhoods, inclusion measures and approximation operators.

consider also costs of searching for relevant approximation spaces in a given family defined by a parameterized approximation space. Any parameterized approximation space $\mathbb{AS}_{\#,\$} = (U, I_\#, \nu_\$)$ is a family of approximation spaces. The cost of searching in such a family for a relevant approximation space for a given concept X approximation can be treated as a factor of the quality measure of approximation of X in $\mathbb{AS}_{\#,\$} = (U, I_\#, \nu_\$)$. Hence, such a quality measure of approximation of X in $\mathbb{AS}_{\#,\$}$ can be defined by

$$Quality_3(\mathbb{AS}_{\#,\$}, X) = h(Quality_2(\mathbb{AS}, X), Cost\_Search(\mathbb{AS}_{\#,\$}, X)), \tag{2.33}$$

where $\mathbb{AS}$ is the result of searching in $\mathbb{AS}_{\#,\$}$, $Cost\_Search(\mathbb{AS}_{\#,\$}, X)$ is the cost of searching in $\mathbb{AS}_{\#,\$}$ for $\mathbb{AS}$, and h is a fusion function, *e.g.*, assuming that the values of $Quality_2(\mathbb{AS}, X)$ and $Cost\_Search(\mathbb{AS}_{\#,\$}, X)$ are normalized to interval $[0,1]$ h could be defined by a linear combination of $Quality_2(\mathbb{AS}, X)$ and $Cost\_Search(\mathbb{AS}_{\#,\$}, X)$ of the form

$$\lambda Quality_2(\mathbb{AS}, X) + (1-\lambda) Cost\_Search(\mathbb{AS}_{\#,\$}, X), \tag{2.34}$$

where $0 \leq \lambda \leq 1$ is a weight measuring an importance of quality and cost in their fusion.

We assume that the fusion functions g, h in the definitions of quality are monotonic relative to each argument.

Let $\mathbb{AS} \in Set\_\mathbb{AS}$ be an approximation space relevant for approximation of $X \subseteq U$, *i.e.*, $\mathbb{AS}$ is the optimal (or semi-optimal) relative to $Quality_2$. By

$$Granulation(\mathbb{AS}_{\#,\$}),$$

we denote a new parameterized approximation space obtained by granulation of $\mathbb{AS}_{\#,\$}$. For example, $Granulation(\mathbb{AS}_{\#,\$})$ can be obtained by reducing the number of attributes or inclusion degrees (*i.e.*, possible values of the inclusion function). Let $\mathbb{AS}'$ be an approximation space in $Granulation(\mathbb{AS}_{\#,\$})$ obtained as the result of searching for optimal (semi-optimal) approximation space in $Granulation(\mathbb{AS}_{\#,\$})$ for approximation of X.

We assume that three conditions are satisfied:

- after granulation of $\mathbb{AS}_{\#,\$}$ to $Granulation(\mathbb{AS}_{\#,\$})$ the following property holds: the cost

$$Cost\_Search(Granulation(\mathbb{AS}_{\#,\$}), X), \tag{2.35}$$

  is much lower than the cost $Cost\_Search(\mathbb{AS}_{\#,\$}, X)$;
- $description(\mathbb{AS}', X)$ is much shorter than $description(\mathbb{AS}, X)$, *i.e.*, the description length of X in the approximation space $\mathbb{AS}'$ is much shorter than the description length of X in the approximation space $\mathbb{AS}$;
- $Quality_1(\mathbb{AS}, X)$ and $Quality_1(\mathbb{AS}', X)$ are sufficiently close.

The last two conditions should guarantee that the values

$$Quality_2(\mathbb{A}\mathbb{S},X) \text{ and } Quality_2(\mathbb{A}\mathbb{S}',X)$$

are comparable and this condition together with the first condition about the cost of searching should assure that

$$Quality_3(Granulation(\mathbb{A}\mathbb{S}_{\#,\$},X)) \text{ is much better than } Quality_3(\mathbb{A}\mathbb{S}_{\#,\$},X). \tag{2.36}$$

Certainly, the phrases already mentioned such as much lower, much shorter, and sufficiently close should be further elaborated. The details will be discussed elsewhere.

Taking into account that parameterized approximation spaces are examples of parameterized granules, one can generalize the above example of parameterized approximation space granulation to the case of granulation of parameterized granules.

In the process of searching for (sub-)optimal approximation spaces, different strategies are used. Let us consider an example of such strategies [252]. In the example, $DT = (U,A,d)$ denotes a decision system (a given sample of data), where U is a set of objects, A is a set of attributes and d is a decision. We assume that for any object x, there is accessible only partial information equal to the A-signature of x (object signature, for short), *i.e.*, $Inf_A(x) = \{(a,a(x)) : a \in A\}$ and analogously for any concept there is only given a partial information about this concept by a sample of objects, *e.g.*, in the form of decision system. One can use object signatures as new objects in a new relational structure $\mathscr{R}$. In this relational structure $\mathscr{R}$ are also modeled some relations between object signatures, *e.g.*, defined by the similarities of these object signatures. Discovery of relevant relations on object signatures is an important step in the searching process for relevant approximation spaces. In this way, a class of relational structures representing perception of objects and their parts is constructed. In the next step, we select a language $\mathscr{L}$ of formulas expressing properties over the defined relational structures and we search for relevant formulas in $\mathscr{L}$. The semantics of formulas (*e.g.*, with one free variable) from $\mathscr{L}$ are subsets of object signatures. Observe that each object signature defines a neighborhood of objects from a given sample (*e.g.*, decision system DT) and another set on the whole universe of objects being an extension of U. In this way, each formula from $\mathscr{L}$ defines a family of sets of objects over the sample and also another family of sets over the universe of all objects. Such families can be used to define new neighborhoods of a new approximation space, *e.g.*, by taking unions of the described above families. In the searching process for relevant neighborhoods, we use information encoded in the given sample. More relevant neighborhoods make it possible to define relevant approximation spaces (from the point of view of the optimization criterion). It is worth to mention that often this searching process is even more compound. For example, one can discover several relational structures (not only one, *e.g.*, $\mathscr{R}$ as it was presented before) and formulas over such structures defining different families of neighborhoods from the original approximation space and next fuse them for obtaining one family of neighborhoods or one neighborhood in a new approximation space. This kind of modeling is typical for hierarchical modeling [11], *e.g.*, when we search for a relevant approximation space for objects composed from parts for which some relevant approximation spaces have been already found.

2.5.3 Rough Sets and Higher Order Vagueness

In [87], it is stressed that vague concepts should have non-crisp boundaries. In the definition presented in this chapter, the notion of boundary region is defined as a crisp set $\mathsf{BN}_B(X)$. However, let us observe that this definition is relative to the subjective knowledge expressed by attributes from B. Different sources of information may use different sets of attributes for concept approximation. Hence, the boundary region can change when we consider these different views. Another aspect is discussed in [232, 250] where it is assumed that information about concepts is incomplete, e.g., the concepts are given only on samples (see, e.g., [77, 89, 115]). From [232, 250] it follows that vague concepts cannot be approximated with satisfactory quality by *static* constructs such as induced membership inclusion functions, approximations or models derived, e.g., from a sample. Understanding of vague concepts can be only realized in a process in which the induced models are adaptively matching the concepts in a dynamically changing environment. This conclusion seems to have important consequences for further development of rough set theory in combination with fuzzy sets and other soft computing paradigms for adaptive approximate reasoning.

2.6 Information Granulation

Information granulation can be viewed as a human way of achieving data compression and it plays a key role in the implementation of the strategy of divide-and-conquer in human problem-solving [318, 176]. Objects obtained as the result of granulation are information granules. Examples of elementary information granules are indiscernibility or tolerance (similarity) classes (see, e.g., [175]). In reasoning about data and knowledge under uncertainty and imprecision many other more compound information granules are used (see, e.g., [206, 205, 229, 241, 242]). Examples of such granules are decision rules, sets of decision rules or classifiers. More compound information granules are defined by means of less compound ones. Note that inclusion or closeness measures between information granules should be considered rather than their strict equality. Such measures are also defined recursively for information granules.

Let us discuss shortly an example of information granulation in the process of modeling patterns for compound concept approximation (see, e.g., [16, 17, 19, 20, 22, 133, 253, 246, 256, 255], [273]. We start from a generalization of information systems. For any attribute $a \in A$ of an information system (U,A) we consider together with the value set V_a of a a relational structure $\mathscr{R}_a$ over the universe V_a (see, e.g., [252]). We also consider a language $\mathscr{L}_a$ of formulas (of the same relational signature as $\mathscr{R}_a$). Such formulas interpreted over $\mathscr{R}_a$ define subsets of Cartesian products of V_a. For example, any formula α with one free variable defines a subset $\|\alpha\|_{\mathscr{R}_a}$

of V_a. Let us observe that the relational structure $\mathscr{R}_a$ (without functions) induces a relational structure over U. Indeed, for any k-ary relation r from $\mathscr{R}_a$ one can define a k-ary relation $g_a \subseteq U^k$ by $(x_1,\ldots,x_k) \in g_a$ if and only if $(a(x_1),\ldots,a(x_k)) \in r$ for any $(x_1,\ldots,x_k) \in U^k$. Hence, one can consider any formula from $\mathscr{L}_a$ as a constructive method of defining a subset of the universe U with a structure induced by $\mathscr{R}_a$. Any such a structure is a new information granule. On the next level of hierarchical modeling, i.e., in constructing new information systems we use such structures as objects and attributes are properties of such structures. Next, one can consider similarity between new constructed objects and then their similarity neighborhoods will correspond to clusters of relational structures. This process is usually more complex. This is because instead of relational structure $\mathscr{R}_a$ we usually consider a fusion of relational structures corresponding to some attributes from A. The fusion makes it possible to describe constraints that should hold between parts obtained by composition from less compound parts. Examples of relational structures can be defined by indiscernibility, similarity, intervals obtained in discretization or symbolic value grouping, preference or spatio-temporal relations (see, e.g., [62, 89, 240]). One can see that parameters to be tuned in searching for relevant[12] patterns over new information systems are, among others, relational structures over value sets, the language of formulas defining parts, and constraints.

2.7 Ontological Framework for Approximation

In a number of papers (see, e.g., [11, 149, 243, 12]), the problem of ontology approximation has been discussed together with possible applications to approximation of compound concepts or to knowledge transfer (see, e.g., [11, 14, 136, 223, 231, 243, 12]). For software RoughICE supporting ontology approximation the reader is referred to the system homepage[13].

In the ontology [270] (vague) concepts and local dependencies between them are specified. Global dependencies can be derived from local dependencies. Such derivations can be used as hints in searching for relevant compound patterns (information granules) in approximation of more compound concepts from the ontology. The ontology approximation problem is one of the fundamental problems related to approximate reasoning in distributed environments. One should construct (in a given language that is different from the ontology specification language) not only approximations of concepts from ontology but also vague dependencies specified in the ontology. It is worthwhile to mention that an ontology approximation should be induced on the basis of incomplete information about concepts and dependencies specified in the ontology. Information granule calculi based on rough sets have been proposed as tools making it possible to solve this problem. Vague dependencies have vague concepts in premisses and conclusions. The approach to approximation of vague dependencies based only on degrees of closeness of concepts from dependencies and their approximations (classifiers) is not satisfactory for approximate

[12] For target concept approximation.

[13] `http://www.mimuw.edu.pl/~bazan/roughice/`

reasoning. Hence, more advanced approach should be developed. Approximation of any vague dependency is a method which allows for any object to compute the arguments "for" and "against" its membership to the dependency conclusion on the basis of the analogous arguments relative to the dependency premisses. Any argument is a compound information granule (compound pattern). Arguments are fused by local schemes (production rules) discovered from data. Further fusions are possible through composition of local schemes, called approximate reasoning schemes (AR schemes) (see, e.g., [20, 153, 205]). To estimate the degree to which (at least) an object belongs to concepts from ontology the arguments "for" and "against" those concepts are collected and next a conflict resolution strategy is applied to them to predict the degree.

2.8 Rough Sets, Approximate Boolean Reasoning and Scalability

Mining large data sets is one of the biggest challenges in KDD. In many practical applications, there is a need of data mining algorithms running on terminals of a client–server database system where the only access to database (located in the server) is enabled by SQL queries.

Unfortunately, the proposed so far data mining methods based on rough sets and Boolean reasoning approach are characterized by high computational complexity and their straightforward implementations are not applicable for large data sets. The critical factor for time complexity of algorithms solving the discussed problem is the number of simple SQL queries like

```
SELECT COUNT FROM aTable WHERE aCondition
```

In this section, we present some efficient modifications of these methods to solve out this problem. We consider the following issues:

- Searching for short reducts from large data sets;
- Searching for best partitions defined by cuts on continuous attributes.

2.8.1 Reduct Calculation

Let us again illustrate the idea of reduct calculation using discernibility matrix (Table 2.1).

Example 2.3. Let us consider the "weather" problem specified by decision system which is represented by decision table (see Table 2.1). Objects are described by four condition attributes and are divided into 2 classes. Let us consider the first 12 observations. In this example, $U = \{1,2,\ldots,12\}$, $A = \{a_1,a_2,a_3,a_4\}$, $CLASS_{no} = \{1,2,6,8\}$, $CLASS_{yes} = \{3,4,5,7,9,10,11,12\}$.

The discernibility matrix can be treated as a board containing $n \times n$ boxes. Noteworthy is the fact that discernibility matrix is symmetrical with respect to the main

Table 2.1. The exemplary "weather" decision table (left) and the compact form of discernibility matrix (right)

date	outlook	temperature	humidity	windy	play
ID	a_1	a_2	a_3	a_4	dec
1	sunny	hot	high	FALSE	no
2	sunny	hot	high	TRUE	no
3	overcast	hot	high	FALSE	yes
4	rainy	mild	high	FALSE	yes
5	rainy	cool	normal	FALSE	yes
6	rainy	cool	normal	TRUE	no
7	overcast	cool	normal	TRUE	yes
8	sunny	mild	high	FALSE	no
9	sunny	cool	normal	FALSE	yes
10	rainy	mild	normal	FALSE	yes
11	sunny	mild	normal	TRUE	yes
12	overcast	mild	high	TRUE	yes

$\mathcal{M}$	1	2	6	8
3	a_1	a_1,a_4	a_1,a_2,a_3,a_4	a_1,a_2
4	a_1,a_2	a_1,a_2,a_4	a_2,a_3,a_4	a_1
5	a_1,a_2,a_3	a_1,a_2,a_3,a_4	a_4	a_1,a_2,a_3
7	a_1,a_2,a_3,a_4	a_1,a_2,a_3	a_1	a_1,a_2,a_3,a_4
9	a_2,a_3	a_2,a_3,a_4	a_1,a_4	a_2,a_3
10	a_1,a_2,a_3	a_1,a_2,a_3,a_4	a_2,a_4	a_1,a_3
11	a_2,a_3,a_4	a_2,a_3	a_1,a_2	a_3,a_4
12	a_1,a_2,a_4	a_1,a_2	a_1,a_2,a_3	a_1,a_4

diagonal, because $M_{i,j} = M_{j,i}$, and that sorting all objects according to their decision classes causes a shift off all empty boxes nearby to the main diagonal. In case of decision table with two decision classes, the discernibility matrix can be rewritten in a more compact form as shown in Table 2.1. The discernibility function is constructed from discernibility matrix by taking a conjunction of all discernibility clauses in which any attribute a_i is substituted by the corresponding Boolean variable x_i. After reducing of all repeated clauses we have[14]:

$$\begin{aligned} f(x_1,x_2,x_3,x_4) =&(x_1)(x_1+x_4)(x_1+x_2)(x_1+x_2+x_3+x_4)(x_1+x_2+x_4) \\ &(x_2+x_3+x_4)(x_1+x_2+x_3)(x_4)(x_2+x_3)(x_2+x_4) \\ &(x_1+x_3)(x_3+x_4)(x_1+x_2+x_4). \end{aligned}$$

One can find relative reducts of the decision table by searching for prime implicants of this discernibility function. The straightforward method allow us to calculate all prime implicants by transformation of the formula to the DNF form (using absorbtion rule $p(p+q) \equiv p$ and other rules for Boolean algebra). One can do it as follows:

$$f = (x_1)(x_4)(x_2+x_3) = x_1x_4x_2 + x_1x_4x_3$$

Thus we have 2 reducts: $R_1 = \{a_1,a_2,a_4\}$ and $R_2 = \{a_1,a_3,a_4\}$.

Every heuristic algorithm for the prime implicant problem can be applied to the discernibility function to solve the minimal reduct problem. One of such heuristics was proposed in [238] and was based on the idea of greedy algorithm, where each attribute is evaluated by its discernibility measure, i.e., the number of pairs of objects which are discerned by the attribute, or, equivalently, the number of its occurrences in the discernibility matrix.

[14] In the formulas $+$ denotes logical disjunction $\vee$ and we omit the conjunction sign $\wedge$ if is this not lead to misunderstanding.

- First we have to calculate the number of occurrences of each attributes in the discernibility matrix:

$$eval(a_1) = disc_{dec}(a_1) = 23, \qquad eval(a_2) = disc_{dec}(a_2) = 23,$$
$$eval(a_3) = disc_{dec}(a_3) = 18, \qquad eval(a_4) = disc_{dec}(a_4) = 16.$$

  Thus a_1 and a_2 are the two most preferred attributes.
- Assume that we select a_1. Now we are taking under consideration only those cells of the discernibility matrix which are not containing a_1. There are 9 such cells only, and the number of occurrences are as the following:

$$eval(a_2) = disc_{dec}(a_1,a_2) - disc_{dec}(a_1) = 7,$$
$$eval(a_3) = disc_{dec}(a_1,a_3) - disc_{dec}(a_1) = 7,$$
$$eval(a_4) = disc_{dec}(a_1,a_4) - disc_{dec}(a_1) = 6.$$

- If this time we select a_2, then the are only 2 remaining cells, and, both are containing a_4;
- Therefore, the greedy algorithm returns the set $\{a_1,a_2,a_4\}$ as a reduct of sufficiently small size.

There is another reason for choosing a_1 and a_4, because they are *core attributes*[15]. One can check that an attribute is a core attribute if and only if occurs in the discernibility matrix as a singleton [238]. Therefore, core attributes can be recognized by searching for all singleton cells of the discernibility matrix. The pseudo-code of this algorithm is presented in Algorithm 1.

The reader may have a feeling that the greedy algorithm for reduct problem has quite a high complexity, because two main operations:

- $disc(B)$ – number of pairs of objects discerned by attributes from B;
- $isCore(a)$ – check whether a is a core attribute;

are defined by the discernibility matrix which is a complex data structure containing $O(n^2)$ cells, and each cell can contain up to $O(m)$ attributes, where n is the number of objects and m is the number of attributes of the given decision table. This suggests that the two main operations need at least $O(mn^2)$ computational time.

Fortunately, both operations can be performed more efficiently. It has been shown [134] that both operations can be calculated in time $O(mn\log n)$ without the necessity to store the discernibility matrix. We present an effective implementation of this heuristics that can be applied to large data sets.

Let $\mathbb{A} = (U,A,dec)$ be a decision system. By a "*counting table*" of a set of objects $X \subset U$ we denote the vector:

$$CountTable(X) = \langle n_1, \ldots, n_d \rangle,$$

where $n_k = card(X \cap CLASS_k)$ is the number of objects from X belonging to the kth decision class.

15 An attribute is called core attribute if and only if it occurs in every reduct [169, 175].

Algorithm 1. Searching for short reduct

```
begin
    B := ∅;
    // Step 1. Initializing B by core attributes
    for a ∈ A do
        if isCore(a) then
            B := B ∪ {a};
        end
    end
    // Step 2. Including attributes to B
    repeat
        a_max := argmax_{a∈A−B} disc_dec(B ∪ {a});
        eval(a_max) := disc_dec(B ∪ {a_max}) − disc_dec(B);
        if (eval(a_max) > 0) then
            B := B ∪ {a};
        end
    until (eval(a_max) == 0) OR (B == A);
    // Step 3. Elimination
    for a ∈ B do
        if (disc_dec(B) = disc_dec(B − {a})) then
            B := B − {a};
        end
    end
end
```

We define a conflict measure of X by

$$conflict(X) = \sum_{i<j} n_i n_j = \frac{1}{2}\left[\left(\sum_{k=1}^{d} n_k\right)^2 - \sum_{k=1}^{d} n_k^2\right].$$

In other words, $conflict(X)$ is the number of pairs $(x,y) \in X \times X$ of objects from different decision classes.

By a *counting table* of a set of attributes B we mean the two-dimensional array $Count(B) = [n_{v,k}]_{v \in INF(B), k \in V_{dec}}$, where

$$n_{v,k} = card(\{x \in U : inf_B(x) = v \text{ and } dec(x) = k\}).$$

Thus $Count(B)$ is a collection of counting tables of equivalence classes of the indiscernibility relation $IND(B)$. It is clear that the complexity time for the construction of counting table is $O(nd \log n)$, where n is the number of objects and d is the number of decision classes. One can also observe that counting tables can be easily constructed in data base management systems using simple SQL queries.

For a given counting table, one can easily calculate the discernibility measure relative to a set of attributes B by

$$disc_{dec}(B) = \frac{1}{2} \sum_{v \neq v', k \neq k''} n_{v,k} \cdot n_{v',k'}.$$

The disadvantage of this equation relates to the fact that it requires $O(S^2)$ operations, where S is the size of the counting table $Count(B)$.

The discernibility measure can be understood as a number of unresolved (by the set of attributes B) conflicts. One can show that:

$$disc_{dec}(B) = conflict(U) - \sum_{[x] \in U/IND(B)} conflict([x]_{IND(B)}). \tag{2.37}$$

Thus, the discernibility measure can be determined in $O(S)$ time:

$$disc_{dec}(B) = \frac{1}{2}\left(n^2 - \sum_{k=1}^{d} n_k^2\right) - \frac{1}{2}\sum_{v \in INF(B)} \left[\left(\sum_{k=1}^{d} n_{v,k}\right)^2 - \sum_{k=1}^{d} n_{v,k}^2\right], \tag{2.38}$$

where $n_k = |CLASS_k| = \sum_v n_{v,k}$ is the size of kth decision class.

Moreover, one can show that attribute a is a core attribute of decision system $\mathbb{A} = (U, A, dec)$ if and only if

$$disc_{dec}(A - \{a\}) < disc_{dec}(A).$$

Thus both operations $disc_{dec}(B)$ and $isCore(a)$ can be performed in linear time with respect to the counting table size.

Example 2.4. In the discussed example, the counting table for a_1 is as follows:

$Count(a_1)$	$dec = no$	$dec = yes$
$a_1 = sunny$	3	2
$a_1 = overcast$	0	3
$a_1 = rainy$	1	3

We illustrate Eqn. (2.38) by inserting some additional columns to the counting table:

$Count(a_1)$	$dec = no$	$dec = yes$	Σ	$conflict(.)$
$a_1 = sunny$	3	2	5	$\frac{1}{2}(5^2 - 2^2 - 3^2) = 6$
$a_1 = overcast$	0	3	3	$\frac{1}{2}(3^2 - 0^2 - 3^2) = 0$
$a_1 = rainy$	1	3	4	$\frac{1}{2}(4^2 - 1^2 - 3^2) = 3$
U	4	8	12	$\frac{1}{2}(12^2 - 8^2 - 4^2) = 32$

Thus $disc_{dec}(a_1) = 32 - 6 - 0 - 3 = 23$.

2.8.2 Discretization of Large Data Sets Stored in Relational Databases

In this section (see [128, 125, 126]), we discuss an application of approximate Boolean reasoning to efficient searching for cuts in large data sets stored in relational databases. Searching for relevant cuts is based on simple statistics which can be efficiently extracted from relational databases. This additional statistical knowledge is making it possible to perform the searching based on Boolean reasoning much more efficient. It can be shown that the extracted cuts by using such reasoning are quite close to optimal.

Searching algorithms for optimal partitions of real-valued attributes, defined by cuts, have been intensively studied. The main goal of such algorithms is to discover cuts which can be used to synthesize decision trees or decision rules of high quality with respect to some quality measures (e.g., quality of classification of new unseen objects, quality defined by the decision tree height, support and confidence of decision rules).

In general, all those problems are hard from computational point of view (e.g., the searching problem for minimal and consistent set of cuts is NP-hard). In consequence, numerous heuristics have been developed for approximate solutions of these problems. These heuristics are based on approximate measures estimating the quality of extracted cuts. Among such measures *discernibility measures* are relevant for the rough set approach.

We outline an approach for solution of a searching problem for optimal partition of real-valued attributes by cuts, assuming that the large data table is represented in a relational database. In such a case, even the linear time complexity with respect to the number of cuts is not acceptable because of the time needed for one step. The critical factor for time complexity of algorithms solving that problem is the number of SQL queries of the form

```
SELECT COUNT
  FROM a Table
     WHERE (an attribute BETWEEN value1 AND value2)
           AND (additional condition)
```

necessary to construct partitions of real-valued attribute sets. We assume the answer time for such queries does not depend on the interval length[16]. Using a straightforward approach to optimal partition selection (with respect to a given measure), the number of necessary queries is of order $O(N)$, where N is the number of preassumed cuts. By introducing some optimization measures, it is possible to reduce the size of searching space. Moreover, using only $O(\log N)$ simple queries, suffices to construct a partition very close to optimal.

Let $\mathbb{A} = (U, A, d)$ be a decision system with real-valued condition attributes. Any cut (a, c), where $a \in A$ and c is a real number, defines two disjoint sets given by

$$U_L(a,c) = \{x \in U : a(x) \leq c\},$$
$$U_R(a,c) = \{x \in U : a(x) > c\}.$$

If both $U_L(a,c)$ and $U_R(a,c)$ are non-empty, then c is called a *cut on attribute a*. The cut (a,c) discerns a pair of objects x, y if either $a(x) < c \leq a(y)$ or $a(y) < c \leq a(x)$.

[16] This assumption is satisfied in some existing database management systems.

Let $\mathbb{A} = (U, A, d)$ be a decision system with real-valued condition attributes and decision classes X_i, for $i = 1, \ldots, r(d)$. A *quality of a cut* (a, c), denoted by $W(a, c)$, is defined by

$$W(a,c) = \sum_{i \neq j}^{r(d)} L_i(a,c) * R_j(a,c) \tag{2.39}$$
$$= \left(\sum_{i=1}^{r(d)} L_i(a,c) \right) * \left(\sum_{i=1}^{r(d)} R_i(a,c) \right) - \sum_{i=1}^{r(d)} L_i(a,c) * R_i(a,c),$$

where $L_i(a,c) = card(X_i \cap U_L(a,c))$ and $R_i(a,c) = card(X_i \cap U_R(a,c))$, for $i = 1, \ldots, r(d)$.

We will be interested in finding cuts maximizing the function $W(a,c)$.

The following definition will be useful. Let $\mathscr{C}_a = \{(a, c_1), \ldots, (a, c_N)\}$ be a set of cuts on attribute a, over a decision system $\mathbb{A}$ and assume $c_1 < c_2 \ldots < c_N$. By a *median of the ith decision class*, denoted by $Median(i)$, we mean the minimal index j for which the cut $(a, c_j) \in \mathscr{C}_a$ minimizes the value $|L_i(a,c_j) - R_i(a,c_j)|$,[17] where L_i and R_i are defined before.

One can use only $O(r(d) * \log N)$ SQL queries to determine the medians of decision classes by using the well-known binary search algorithm.

Then one can show that the quality function $W_a(i) \stackrel{\text{def}}{=} W(a, c_i)$, for $i = 1, \ldots, N$, is increasing in $\{1, \ldots, min\}$ and decreasing in $\{max, \ldots, N\}$, where *min* and *max* are defined by

$$min = \min_{1 \le i \le N} Median(i),$$
$$max = \max_{1 \le i \le N} Median(i).$$

In consequence, the search space for maximum of $W(a, c_i)$ is reduced to $i \in [min, max]$.

Now, one can apply the divide and conquer strategy to determine the best cut, given by $c_{Best} \in [c_{min}, c_{max}]$, with respect to the chosen quality function. First, we divide the interval containing all possible cuts into k intervals. Using some heuristics, one then predict the interval which most probably contains the best cut. This process is recursively applied to that interval, until the considered interval consists of one cut. The problem which remains to be solved is how to define such approximate measures which could help us to predict the suitable interval.

Let us consider a simple probabilistic model. Let (a, c_L), (a, c_R) be two cuts such that $c_L < c_R$ and $i = 1, \ldots, r(d)$. For any cut (a, c) satisfying $c_L < c < c_R$, we assume that $x_1, \ldots, x_{r(d)}$, where $x_i = card(X_i \cap U_L(a,c) \cap U_R(a,c))$ are independent random variables with uniform distribution over sets $\{0, \ldots, M_1\}$, ..., $\{0, \ldots, M_{r(d)}\}$, respectively, that

$$M_i = M_i(a, c_L, c_R) = card(X_i \cap U_L(a, c_R) \cap U_R(a, c_L)).$$

[17] The minimization means that $|L_i(a,c_j) - R_i(a,c_j)| = \min_{1 \le k \le N} |L_i(a,c_k) - R_i(a,c_k)|$.

Under these assumptions the following fact holds. For any cut $c \in [c_L, c_R]$, the mean $E(W(a,c))$ of quality $W(a,c)$, is given by

$$E(W(a,c)) = \frac{W(a,c_L) + W(a,c_R) + conflict((a,c_L),(a,c_R))}{2}, \tag{2.40}$$

where $conflict((a,c_L),(a,c_R)) = \sum_{i \neq j} M_i * M_j$.

In addition, the standard deviation of $W(a,c)$ is given by

$$D^2(W(a,c)) = \sum_{i=1}^{n} \left[\frac{M_i(M_i+2)}{12} \left(\sum_{j \neq i} (R_j(a,c_R) - L_j(a,c_L)) \right)^2 \right]. \tag{2.41}$$

Formulas (2.40) and (2.41) can be used to construct a predicting measure for the quality of the interval $[c_L, c_R]$:

$$Eval([c_L, c_R], \alpha) = E(W(a,c)) + \alpha \sqrt{D^2(W(a,c))}, \tag{2.42}$$

where the real parameter $\alpha \in [0,1]$ can be tuned in a learning process.

To determine the value $Eval([c_L, c_R], \alpha)$, we need to compute the numbers

$$L_1(a,c_L), \ldots, L_{r(d)}(a,c_L), M_1, \ldots, M_{r(d)}, R_1(a,c_R), \ldots, R_{r(d)}(a,c_R).$$

This requires $O(r(d))$ SQL queries of the form

```
SELECT COUNT
FROM DecTable
WHERE (attribute a BETWEEN value1 AND value2)
      AND (dec = i).
```

Hence, the number of queries required for running this algorithm is

$$O(r(d) k \log_k N).$$

In practice, we set $k = 3$, since the function $f(k) = r(d) k \log_k N$ over positive integers is taking minimum for $k = 3$.

Numerous experiments on different data sets have shown that the proposed solution allows one to find a cut which is very close to the optimal one. For more details the reader is referred to the literature (see [125, 126]).

2.9 Rough Sets and Logic

The father of contemporary logic is a German mathematician Gottlob Frege (1848-1925). He thought that mathematics should not be based on the notion of set but on the notions of logic. He created the first axiomatized logical system but it was not understood by the logicians of those days.

During the first three decades of the 20th century, there was a rapid development in logic bolstered to a great extent by Polish logicians, especially Alfred Tarski (1901-1983) (see,e.g., [287]).

Development of computers and their applications stimulated logical research and widened their scope.

When we speak about logic, we generally mean *deductive logic*. It gives us tools designed for deriving true propositions from other true propositions. Deductive reasoning always leads to true conclusions. The theory of deduction has well established generally accepted theoretical foundations. Deductive reasoning is the main tool used in mathematical reasoning and found no application beyond it.

Rough set theory has contributed to some extent to various kinds of deductive reasoning. Particularly, various kinds of logics based on the rough set approach have been investigated, rough set methodology contributed essentially to modal logics, many valued logic, intuitionistic logic and others (see, e.g., [5, 6, 42, 43, 46, 58, 57, 110, 111, 123, 122, 124, 141, 143, 144, 146, 167, 168, 197, 198, 209, 210, 211, 212, 213, 214, 215, 296, 295, 297, 298]).

A summary of this research can be found in [196, 32] and interested reader is advised to consult these volumes.

In natural sciences (e.g., in physics) *inductive reasoning* is of primary importance. The characteristic feature of such reasoning is that it does not begin from axioms (expressing general knowledge about the reality) like in deductive logic, but some partial knowledge (examples) about the universe of interest are the starting point of this type of reasoning, which are generalized next and they constitute the knowledge about wider reality than the initial one. In contrast to deductive reasoning, inductive reasoning does not lead to true conclusions but only to probable (possible) ones. Also in contrast to the logic of deduction, the logic of induction does not have uniform, generally accepted, theoretical foundations as yet, although many important and interesting results have been obtained, e.g., concerning statistical and computational learning and others.

Verification of validity of hypotheses in the logic of induction is based on experiment rather than the formal reasoning of the logic of deduction. Physics is the best illustration of this fact.

The research on inductive logic has a few centuries' long history and outstanding English philosopher John Stuart Mill (1806-1873) is considered its father [114].

The creation of computers and their innovative applications essentially contributed to the rapid growth of interest in inductive reasoning. This domain develops very dynamically thanks to computer science. Machine learning, knowledge discovery, reasoning from data, expert systems and others are examples of new directions in inductive reasoning. It seems that rough set theory is very well suited as a theoretical basis for inductive reasoning. Basic concepts of this theory fit very well to represent and analyze knowledge acquired from examples, which can be next used as starting point for generalization. Besides, in fact rough set theory has been successfully applied in many domains to find patterns in data (data mining) and acquire knowledge from examples (learning from examples). Thus, rough set theory seems to be another candidate as a mathematical foundation of inductive reasoning [19, 133, 251].

The most interesting from computer science point of view is *common sense* reasoning. We use this kind of reasoning in our everyday life, and examples of such kind of reasoning we face in news papers, radio TV etc., in political, economic etc., debates and discussions.

The starting point to such reasoning is the knowledge possessed by the specific group of people (*common knowledge*) concerning some subject and intuitive methods of deriving conclusions from it. We do not have here possibilities of resolving the dispute by means of methods given by deductive logic (reasoning) or by inductive logic (experiment). So the best known methods of solving the dilemma is voting, negotiations or even war. See e.g., Gulliver's Travels [282], where the hatred between Tramecksan (High-Heels) and Slamecksan (Low-Heels) or disputes between Big-Endians and Small-Endians could not be resolved without a war.

These methods do not reveal the truth or falsity of the thesis under consideration at all. Of course, such methods are not acceptable in mathematics or physics. Nobody is going to solve by voting, negotiations or declare a war – the truth of Fermat's theorem or Newton's laws.

Reasoning of this kind is the least studied from the theoretical point of view and its structure is not sufficiently understood, in spite of many interesting theoretical research in this domain [51]. The meaning of common sense reasoning, considering its scope and significance for some domains, is fundamental and rough set theory can also play an important role in it but more fundamental research must be done to this end [237].

In particular, the rough truth introduced in [167] and studied, e.g., in [6] seems to be important for investigating commonsense reasoning in the rough set framework.

Let us consider a simple example. In the considered decision system we assume $U = Birds$ is a set of birds that are described by some condition attributes from a set A. The decision attribute is a binary attribute $Files$ with possible values *yes* if the given bird flies and *no*, otherwise. Then, we define (relative to an information system $\mathbb{A} = (U, A)$) the set of abnormal birds by $Ab_A(Birds) = \mathsf{LOW}_A(\{x \in Birds : Flies(x) = no\})$. Hence, we have, $Ab_A(Birds) = Birds - \mathsf{UPP}_A(\{x \in Birds : Flies(x) = yes\})$ and $Birds - Ab_A(Birds) = \mathsf{UPP}_A(\{x \in Birds : Flies(x) = yes\})$. It means that for normal birds it is consistent, with knowledge represented by A, to assume that they can fly, i.e., it is possible that they can fly. One can optimize $Ab_A(Birds)$ using A to obtain minimal boundary region in the approximation of $\{x \in Birds : Flies(x) = no\}$.

It is worthwhile to mention that in [38] has been presented an approach combining the rough sets with nonmonotonic reasoning. There are distinguished some basic concepts that can be approximated on the basis of sensor measurements and more complex concepts that are approximated using so called transducers defined by first order theories constructed overs approximated concepts. Another approach to commonsense reasoning has been developed in a number of papers (see, e.g., [19, 133, 153, 206, 237]). The approach is based on an ontological framework for approximation. In this approach approximations are constructed for concepts and dependencies between the concepts represented in a given ontology expressed, e.g., in natural language. Still another approach combining rough sets with logic programming is discussed in [301].

To recapitulate, the characteristics of the three above mentioned kinds of reasoning are given below:

1. deductive:
   - reasoning method: axioms and rules of inference;
   - applications: mathematics;
   - theoretical foundations: complete theory;
   - conclusions: true conclusions from true premisses;
   - hypotheses verification: formal proof.
2. inductive:
   - reasoning method: generalization from examples;
   - applications: natural sciences (physics);
   - theoretical foundation: lack of generally accepted theory;
   - conclusions: not true but probable (possible);
   - hypotheses verification - experiment.
3. common sense:
   - reasoning method based on common sense knowledge with intuitive rules of inference expressed in natural language;
   - applications: every day life, humanities;
   - theoretical foundation: lack of generally accepted theory;
   - conclusions obtained by mixture of deductive and inductive reasoning based on concepts expressed in natural language, e.g., with application of different inductive strategies for conflict resolution (such as voting, negotiations, cooperation, war) based on human behavioral patterns;
   - hypotheses verification - human behavior.

There are numerous issues related to approximate reasoning under uncertainty. These issues are discussed in books on granular computing, rough mereology, and computational complexity of algorithmic problems related to these issues. For more detail, the reader is referred to the following books [176, 118, 36, 121, 199].

Finally, we would like to stress that still much more work should be done to develop approximate reasoning methods for making progress in development intelligent systems. This idea was very well expressed by Professor Leslie Valiant[18]:

> *A fundamental question for artificial intelligence is to characterize the computational building blocks that are necessary for cognition. A specific challenge is to build on the success of machine learning so as to cover broader issues in intelligence.... This requires, in particular a reconciliation between two contradictory characteristics – the apparent logical nature of reasoning and the statistical nature of learning.*

---

[18] `http://people.seas.harvard.edu/~valiant/researchinterests.htm`

2.10 Interactive Rough Granular Computing (IRGC)

There are many real-life problems that are still hard to solve using the existing methodologies and technologies. Among such problems are, e.g., classification and understanding of medical images, control of autonomous systems like unmanned aerial vehicles or robots, and problems related to monitoring or rescue tasks in multiagent systems. All of these problems are closely related to intelligent systems that are more and more widely applied in different real-life projects.

One of the main challenges in developing intelligent systems is discovering methods for approximate reasoning from measurements to perception, i.e., deriving from concepts resulting from sensor measurements concepts or expressions enunciated in natural language that express perception understanding.

Nowadays, new emerging computing paradigms are investigated attempting to make progress in solving problems related to this challenge. Further progress depends on a successful cooperation of specialists from different scientific disciplines such as mathematics, computer science, artificial intelligence, biology, physics, chemistry, bioinformatics, medicine, neuroscience, linguistics, psychology, sociology. In particular, different aspects of reasoning from measurements to perception are investigated in psychology [9, 75], neuroscience [195], layered learning [273], mathematics of learning [195], machine learning, pattern recognition [77], data mining [89] and also by researchers working on recently emerged computing paradigms such as computing with words and perception [318], granular computing [153], rough sets, rough-mereology, and rough-neural computing [153].

One of the main problems investigated in machine learning, pattern recognition [77] and data mining [89] is concept approximation. It is necessary to induce approximations of concepts (models of concepts) from available experimental data. The data models developed so far in such areas like statistical learning, machine learning, pattern recognition are not satisfactory for approximation of compound concepts resulting in the perception process. Researchers from the different areas have recognized the necessity to work on new methods for concept approximation (see, e.g., [29, 300]). The main reason is that these compound concepts are, in a sense, too far from measurements which makes the searching for relevant (for their approximation) features infeasible in a huge space. There are several research directions aiming at overcoming this difficulty. One of them is based on the interdisciplinary research where the results concerning perception in psychology or neuroscience are used to help to deal with compound concepts (see, e.g., [77]). There is a great effort in neuroscience towards understanding the hierarchical structures of neural networks in living organisms [45, 195]. Convolutional networks (ConvNets) which are a biologically inspired trainable architecture that can learn invariant features, were developed (see, e.g., [288]). Also mathematicians are recognizing problems of learning as the main problem of the current century [195]. The problems discussed so far are also closely related to complex system modeling. In such systems again the problem of concept approximation and reasoning about perceptions using concept approximations is one of the challenges nowadays. One should take into account that modeling complex phenomena entails the use of local models (captured by local agents, if one would like to use the multi-agent terminology [79, 319]) that next should be fused [290]. This process involves the negotiations between agents [79] to resolve contradictions and conflicts in local modeling. This kind of modeling will become more and more important in solving complex real-life problems which we are unable to model using traditional analytical approaches. The latter approaches lead to exact models. However, the necessary assumptions used to develop them are causing the resulting solutions to be too far from reality to be accepted. New methods or even a new science should be developed for such modeling [54].

One of the possible solutions in searching for methods for compound concept approximations is the layered learning idea [273]. Inducing concept approximation should be developed hierarchically starting from concepts close to sensor measurements to compound target concepts related to perception. This general idea can be realized using additional domain knowledge represented in natural language. For example, one can use principles of behavior on the roads, expressed in natural language, trying to estimate, from recordings (made, e.g., by camera and other sensors) of situations on the road, if the current situation on the road is safe or not. To solve such a problem one should develop methods for concept approximations together with methods aiming at approximation of reasoning schemes (over such concepts) expressed in natural language. Foundations of such an approach are based on rough set theory [169] and its extension rough mereology [153, 200, 201, 203, 11, 199], both discovered in Poland.

Objects we are dealing with are information granules. Such granules are obtained as the result of information granulation [318]:

> *Information granulation can be viewed as a human way of achieving data compression and it plays a key role in implementation of the strategy of divide-and-conquer in human problem-solving.*

Constructions of information granules should be robust with respect to their input information granule deviations. In this way also a granulation of information granule constructions is considered. As the result we obtain the so called AR schemes (AR networks) [153, 200, 201, 203]. AR schemes can be interpreted as complex patterns [89]. Searching methods for such patterns relevant for a given target concept have been developed [153, 11]. Methods for deriving relevant AR schemes are of high computational complexity. The complexity can be substantially reduced by using domain knowledge. In such a case AR schemes are derived along reasoning schemes in natural language that are retrieved from domain knowledge. Developing methods for deriving such AR schemes is one of the main goals of our projects.

Granulation is a computing paradigm, among others like self-reproduction, self-organization, functioning of brain, Darwinian evolution, group behavior, cell membranes, and morphogenesis, that are abstracted from natural phenomena. Granulation is inherent in human thinking and reasoning processes. Granular computing (GrC) provides an information processing framework where computation and operations are performed on information granules, and it is based on the realization that precision is sometimes expensive and not much meaningful in modeling and controlling complex systems. When a problem involves incomplete, uncertain, and vague information, it may be difficult to differentiate distinct elements and one may find it convenient to consider granules for its handling. The structure of granulation can be often defined using methods based on rough sets, fuzzy sets or their combination. In this consortium, rough sets and fuzzy sets work synergistically, often with other soft computing approaches, and use the principle of granular computing. The developed systems exploit the tolerance for imprecision, uncertainty, approximate reasoning and partial truth under soft computing framework and is capable of achieving tractability, robustness, and close resemblance with human like (natural) decision making for pattern recognition in ambiguous situations [236]. Qualitative reasoning requires to develop methods supporting approximate reasoning under uncertainty about non-crisp concepts, often vague concepts. One of the very general scheme of tasks for such qualitative reasoning can be described as follows. From some basic objects (called in different areas as patterns, granules or molecules) it is required to construct (induce) complex objects satisfying a given specification (often, expressed in natural language specification) to a satisfactory degree. For example, in learning concepts from examples we deal with tasks where a partial information about the specification is given by examples and counter examples concerning of classified objects. As examples of such complex objects one can consider classifiers considered in Machine Learning or Data Mining, new medicine against some viruses or behavioral patterns of cell interaction induced from interaction of biochemical processes realized in cells. Over the years we have learned how to solve some of such tasks while many of them are still challenges. One of the reasons is that the discovery process of complex objects relevant for the given specification requires multilevel reasoning with necessity of discovering on each level the relevant structural objects and their properties. The searching space for such structural objects and properties is very huge and this, in particular, causes that fully automatic methods are not feasible using the exiting computing technologies. However, this process can be supported by domain knowledge used which can be used for generating hints in the searching process (see, e.g., [11]). This view is consistent with [28] (see, page 3 of Foreword):

> *[...] Tomorrow, I believe, every biologist will use computer to define their research strategy and specific aims, manage their experiments, collect their results, interpret their data, incorporate the findings of others, disseminate their observations, and extend their experimental observations - through exploratory discovery and modeling - in directions completely unanticipated.*

Rough sets, discovered by Zdzisław Pawlak [166], and fuzzy sets, due to Lotfi Zadeh [317], separately and in combination have shown quite strong potential for supporting the searching process for the relevant complex objects (granules) discussed above (see, e.g., [154, 153, 173, 11, 148]). Fuzzy set theory addresses gradualness of knowledge, expressed by the fuzzy membership, whereas rough set theory addresses granularity of knowledge, expressed by the indiscernibility relation.

Computations on granules should be interactive. This requirement is fundamental for modeling of complex systems [55]. For example, in [139] this is expressed by the following sentence:

> *[...] interaction is a critical issue in the understanding of complex systems of any sorts: as such, it has emerged in several well-established scientific areas other than computer science, like biology, physics, social and organizational sciences.*

Interactive Rough Granular Computing (IRGC) is an approach for modeling interactive computations (see, e.g., [255]). IRGC are progressing by interactions between granules (structural objects of quite often high order type) discovered from data and domain knowledge. In particular, interactive information systems (IIS) are dynamic granules used for representing the results of the agent interaction with the environments. IIS can be also applied in modeling more advanced forms of interactions such as hierarchical interactions in layered granular networks or generally in hierarchical modeling. The proposed approach [255, 256] is based on rough sets but it can be combined with other soft computing paradigms such as fuzzy sets or evolutionary computing, and also with machine learning and data mining techniques. The notion of the highly interactive granular system is clarified as the system in which intrastep interactions with the external as well as with the internal environments take place. Two kinds of interactive attributes are distinguished: perception attributes, including sensory ones and action attributes.

The outlined research directions in this section create a step toward understanding the nature of reasoning from measurements to perception. These foundations are crucial for constructing intelligent systems for many real-life projects. The recent progress in this direction based on rough sets and granular computing is reported in [255, 256].

In the following section, we outline three important challenging topics.

2.10.1 Context Inducing and IRGC

Reasoning about context belongs to the main problems investigated in AI for many years (see, e.g., [112, 222, 269]). One of the old and still challenging problem in machine learning, pattern recognition and data mining is feature discovery (feature construction, feature extraction) [77]. This problem is related to discovery of structures of objects or contexts in which analyzed objects should be considered. In this section, we discuss an application of information systems for context modeling. The approach is based on fusion of information systems (or decision systems) with constraints. The constraints can be defined by means of relations over sets of attribute values or their Cartesian products. Objects on the next level of modeling are relational structures over signatures (or sets of signatures) of arguments of fusion operation. In this way, one can obtain as objects on higher level of modeling indiscernibility (similarity) classes of objects, time windows, their clusters, sequences of time windows and their sets. Indiscernibility classes over objects representing sequences of time windows are sets of such sequences and they may represent information about processes.

Let us consider one simple example illustrating this approach elaborated, e.g., in [254, 255, 256].

In the process of searching for (sub-)optimal approximation spaces, different strategies may be used. Let us consider an example of such strategy presented in [252]. In this example, $DT = (U, A, d)$ denotes a decision system (a given sample of data), where U is a set of objects, A is a set of attributes and d is a decision. We assume that for any object $x \in U$, only partial information equal to the A-signature of x (object signature, for short) is accessible, i.e., $Inf_A(x) = \{(a, a(x)) : a \in A\}$. Analogously, for any concept there are only given a partial information about this concept by means of a sample of objects, e.g., in the form of decision table. One can use object signatures as new objects in a new relational structure $\mathscr{R}$. In this relational structure $\mathscr{R}$ some relations between object signatures are also modeled, e.g., defined by the similarities of these object signatures (see Figure 2.3).

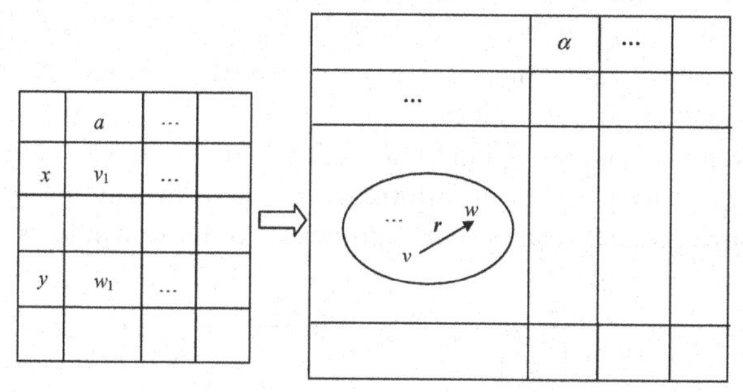

Fig. 2.3. Granulation to tolerance classes. r is a similarity (tolerance) relation defined over signatures of objects.

Discovery of relevant relations between object signatures is an important step in searching for relevant approximation spaces. In this way, a class of relational structures representing perception of objects and their parts is constructed. In the next step, we select a language $\mathscr{L}$ consisting of formulas expressing properties over the defined relational structures and we search for relevant formulas in $\mathscr{L}$. The semantics of formulas (e.g., with one free variable) from $\mathscr{L}$ are subsets of object signatures. Note, that each object signature defines a neighborhood of objects from a given sample (e.g., decision system DT) and another set on the whole universe of objects being an extension of U. Thus, each formula from $\mathscr{L}$ defines a family of sets of objects over the sample and also another family of sets over the universe of all objects. Such families can be used to define new neighborhoods for a new approximation space by, e.g., taking their unions. In the process of searching for relevant neighborhoods, we use information encoded in the available sample. More relevant neighborhoods make it possible to define more relevant approximation spaces (from the point of view of the optimization criterion). Following this scheme, the next level of granulation may be related to clusters of objects (relational structures) for a current level (see Figure 2.4).

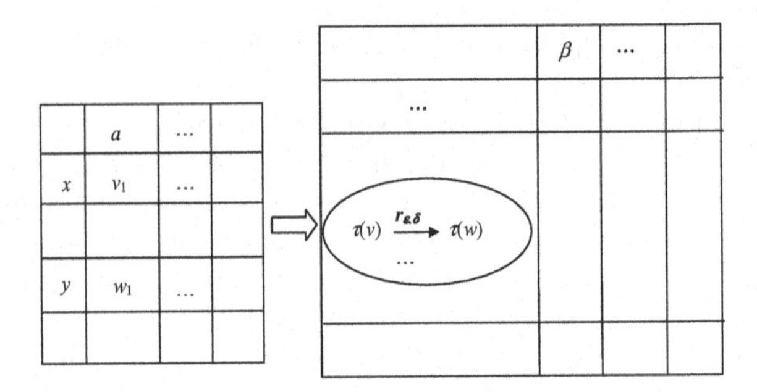

Fig. 2.4. Granulation of tolerance relational structures to clusters of such structures. $r_{\varepsilon,\delta}$ is a relation with parameters ε, δ on similarity (tolerance) classes.

In Figure 2.4 τ denotes a similarity (tolerance) relation on vectors of attribute values, $\tau(v) = \{u: \ v \ \tau \ u\}$, $\tau(v) \ r_{\varepsilon,\delta} \ \tau(w)$ iff $dist(\tau(v), \tau(w)) \in [\varepsilon - \delta, \varepsilon + \delta]$, and $dist(\tau(v), \tau(w)) = inf\{dist(v', w') : \ (v', w') \in \tau(v) \times \tau(w)\}$ where *dist* is a distance function on vectors of attribute values.

One more example is illustrated in Figure 2.5, where the next level of hierarchical modeling is created by defining an information system in which objects are time windows and attributes are (time-related) properties of these windows.

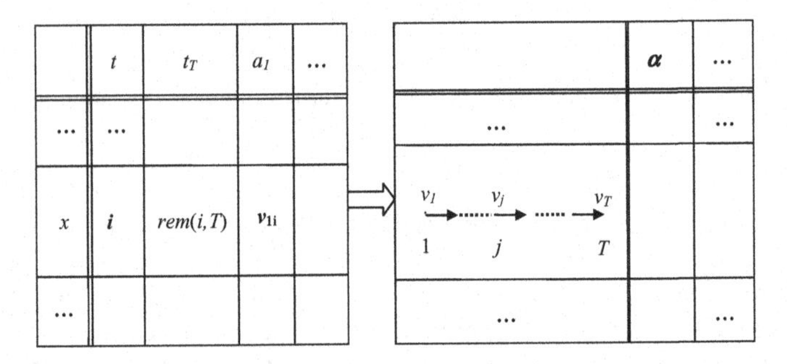

Fig. 2.5. Granulation of time points into time windows. T is the time window length, $v_j = (v_{1j}, \ldots, v_{Tj})$ for $j = 1, \ldots, T$, $rem(i, T)$ is the remainder from division of i by T, α is an attribute defined over time windows.

It is worth mentioning that quite often this searching process is even more sophisticated. For example, one can discover several relational structures (e.g., corresponding to different attributes) and formulas over such structures defining different families of neighborhoods from the original approximation space. As a next step, such families of neighborhoods can be merged into neighborhoods in a new, higher degree approximation space.

The proposed approach is making it possible to construct information systems (or decision system) on a given level of hierarchical modeling from information systems from lower level(s) by using some constraints in joining objects from underlying information systems. In this way, structural objects can be modeled and their properties can be expressed in constructed information systems by selecting relevant attributes. These attributes are defined with use of a language that makes use of attributes of systems from the lower hierarchical level as well as relations used to define constraints. In some sense, the objects on the next level of hierarchical modeling are defined using the syntax from the lover level of the hierarchy. Domain knowledge is used to aid the discovery of relevant attributes (features) on each level of hierarchy. This domain knowledge can be provided, e.g., by concept ontology together with samples of objects illustrating concepts from this ontology. Such knowledge is making it feasible to search for relevant attributes (features) on different levels of hierarchical modeling.

In Figure 2.6 we symbolically illustrate the transfer of knowledge in a particular application. It is a depiction of how the knowledge about outliers in handwritten digit recognition is transferred from expert to a software system. We call this process *knowledge elicitation* [135, 136, 137]. Observe, that the explanations given by expert(s) are expressed using a subset of natural language limited by using concepts from provided ontology only. Concepts from higher levels of ontology are gradually approximated by the system from concepts on lower levels.

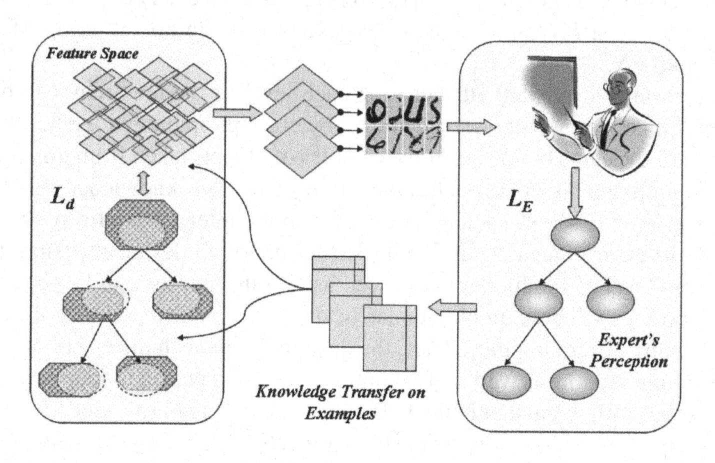

Fig. 2.6. Expert's knowledge elicitation.

This kind of approach is typical for hierarchical modeling [20, 14, 19, 133, 11, 12]. This is, in particular, the case when we search for a relevant approximation space for objects composed from parts for which some approximation spaces, relevant to components, have already been found. We find that hierarchical modeling is required for approximation of complex vague concepts, as in [137, 195].

2.10.2 Process Mining and IRGC

The rapid expansion of the Internet has resulted not only in the ever growing amount of data therein stored, but also in the burgeoning complexity of the concepts and phenomena pertaining to those data. This issue has been vividly compared in [49] to the advances in human mobility from the period of walking afoot to the era of jet travel. These essential changes in data have brought new challenges to the development of new data mining methods, especially that the treatment of these data increasingly involves complex processes that elude classic modeling paradigms. Types of datasets currently regarded "hot", like biomedical, financial or net user behavior data are just a few examples. Mining such temporal or complex data streams is on the agenda of many research centers and companies worldwide (see, e.g., [1, 219]). In the data mining community, there is a rapidly growing interest in developing methods for process mining, e.g., for discovery of structures of temporal processes from observations (recorded data). Works on process mining, e.g., [27, 113, 294, 308, 299] have recently been undertaken by many renowned centers worldwide[19]. This research is also related to functional data analysis (cf. [208]), cognitive networks (cf. [162]), and dynamical system modeling in biology (cf. [47]).

Let us consider an illustrative example explaining motivation for discovery of process models from data.

This problem is illustrated in Figure 2.7. It is assumed that from granules G, G_1, G_2 representing the sets of the paths of the processes, their models in the form of Petri nets PN, PN_1, PN_2, respectively, were induced. Then, the structure of interaction between PN_1 and PN_2 can be described by an operation transforming PN_1, PN_2 into PN.

The discovery of relevant attributes on each level of the hierarchy can be supported by domain knowledge provided e.g., by concept ontology together with the illustration of concepts by means of the samples of objects taken from this concepts and their complements [11]. Such application of domain knowledge often taken from human experts serves as another example of the interaction of a system (classifier) with its environment. Additionally, such support of relevant attributes discovery on given level of the hierarchy, as well as on other levels, can be found using different ontologies. These ontologies can be described by different sets of formulas and possibly by different logics. Thus, the description of such discovery of relevant attributes in interaction, as well as its support give a good reason for applying fibring logics methods [50]. Note that in the hierarchical modeling of relevant complex patterns also top-down interactions of the higher levels of the hierarchy with the lower levels should be considered, e.g., if the patterns constructed on higher levels are not relevant for the target task, the top-down interaction should inform lower levels about the necessity of searching for new patterns.

There are are numerous papers based on the rough set approach on discovery of concurrent processes from data (see, e.g. [248, 247, 249, 275, 191, 276, 192, 155, 278, 156, 157, 158, 159, 281, 280, 283, 279, 160, 116, 161, 119, 117, 120, 277, 129, 36]). In [131, 129] was outlined an approach to discovery of processes from data and domain knowledge which is based on RGC philosophy. This research was

[19] http://www.isle.org/~langley/,
http://soc.web.cse.unsw.edu.au/bibliography/discovery/index.html,
http://www.processmining.org/

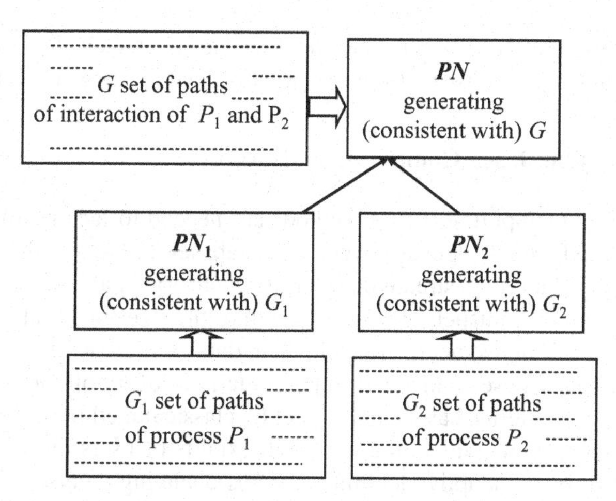

Fig. 2.7. Discovery of interaction structure.

initiated by the idea of Professor Zdzisław Pawlak presented in [170], where data tables are treated as partial specifications of concurrent processes. Rows in a given data table are examples of global states and columns represent local processes. One of the solutions presented in the above papers was based on decomposition of data tables into modules defined by reducts of data tables. The modules are linked by constraints defined by rules extracted from data. In another approach, first from a given data table decision rules are extracted (e.g., a set of minimal decision rules) and such a set of decision rules is used as knowledge encoded in the data table or theory defined by data table. Next, the set of all global states is defined as equal to the maximal set of objects (global states) consistent with the theory. There were proposed methods for automatic generation from a given data table a (colored) Petri net with the reachability set equal to the maximal consistent set of states consistent with the theory generated from the data table. The reader is referred to the Web page `http://rsds.univ.rzeszow.pl` for information on the developed software (ROSECON) for inducing Petri nets from data tables.

An important role in discovering Petri nets play the inhibitory rules (see, e.g., [36]). The reader interested in complexity results related to such rules as well as to consistent extensions of information systems is referred to, e.g., [116, 119, 117, 36, 120].

Here, we would like to formulate some challenges related to discovery of concurrent processes from data tables occurring in hierarchical modeling by using IRGC. On higher levels of hierarchy, the structure of objects becomes complex, e.g., indiscernibility classes of data tables considered on higher level of the hierarchy can be equal to sets of paths of structural states. The theories of such data tables are much more complex than considered before. The rules in such a theory discovered from data may require extension to (spatio-)temporal decision rules or temporal association rules or even more complex rules defined by different temporal logics. The challenges are related to discovery of relevant rules and theories over such rules as well as to inducing, e.g, Petri nets consistent with theories defined by such constraints.

2.10.3 Perception Based Computing and IRGC

Perception Based Computing (PBC) methods are needed to face problems of data mining (DM) and knowledge discovery in databases (KDD) with dynamically evolving complex data (e.g., stream data sources, sensory data). Another challenge, making PBC methods indispensable, is a growth of the size and complexity of data sources (e.g., Web sources, neuro-imaging data, data from network interactions) in open environments. These challenges, in particular, discovery of complex concepts such as behavioral patterns, hardly can be met by classical methods [195]. They can be met by KDD systems which dialogue with experts or users during the discovery process[300] or by adaptive learning systems changing themselves during the learning process as the response to evolving data.

Another area where PBC methods are needed is a multi-agent systems field. Behavior steering and coordination of multi-agent coalitions acting and cooperating in open, unpredictable environments call for interactive algorithms [56], i.e. algorithms interacting with the environment during performing particular steps of computations or changing themselves during the process of computation. Next challenge of this type comes from human – robot interaction. The problem of human control over autonomously coordinating swarms of robots is the central challenge in this field which should be solved before human – robot teams can be taken out of laboratories and put to practical use.

Coordination and control are essentially perception based thus PBC methods are indispensable for designing and behavior description of cognitive systems and for understanding interactions in layered granular networks [176], where granules can be interpreted both as data patterns and agents (e.g., robots or movable sensors). Granules in such networks, which are additionally self-organizing, can be also understood as cores in pertinent multi-core computing engines in structurally and runtime reconfigurable hardware, what makes PBCs useful in computer engineering as well as an essential part of cognitive informatics.

Current works are aimed at developing methods based on the generalized information systems (a special kind of data tables) and the rough set approach for representing partial information on interactions in layered granular networks [82, 84, 83], [255, 256]. The idea of the representation of interactions using information systems has some roots in such approaches as rough sets introduced by Zdzisław Pawlak [166], the information flow by Jon Barwise [10] or Chu spaces [8], `http://chu.stanford.edu/`. Information systems are used to represent granules of different complexity and the interactions among them [255, 83]. Rough sets are used for vague concept approximation [175], for example, in the approximation of ontologies given by experts (see, e.g., [11]).

Perception based computing provides capability to compute and reason with perception based information as humans do to perform a wide variety of physical and mental tasks without any measurement and computation. Reflecting the finite ability of the sensory organs and (finally the brain) to resolve details, perceptions are inherently granular. Boundaries of perceived granules (e.g., classes) are unsharp and the values of the attributes they can take are granulated. In general, perception may be considered as understanding of sensory information. This point of view is, e.g., presented in Computing with Words and Perception [318] which

> *derives from the fact that it opens the door to computation and reasoning with information which is perception – rather than measurement-based. Perceptions play a key role in human cognition, and underlie the remarkable human capability to perform a wide variety of physical and mental tasks without any measurements and any computations. Everyday examples of such tasks are driving a car in city traffic, playing tennis and summarizing a story.*

The need for perception based computing appears, for example, in problems of analysis of complex processes that result from the interaction of many component processes and from control over such process. A component process control is aimed at achieving the desired patterns of the system behaviors. This task is a challenge for areas such as multi-agent systems or autonomous systems [284, 228, 106]. Perceived properties of complex processes are often complex vague concepts, about which only partial information is available. Also information about the satisfiability of such concepts determines activating complex actions. It is worth noting that actions can be initiated at different levels of the hierarchy of concepts from a given ontology and that a prediction of action activation at higher levels of the hierarchy is usually conditioned by the perception of properties depending on the history of computations in which information about actions conducted also on lower levels of the hierarchy and their results is stored. Discovery of search strategies for new essential properties at higher levels of hierarchy becomes a challenge and is crucial for understanding of perception. The values of these attributes depend on the history of computing (with registered information about the actions performed on the actual and the lower levels and their results). These new features determine the perception of satisfiability degrees of complex concepts mentioned above conditioning execution of actions on the considered level of hierarchy. The difficulties of analysis and synthesis of perceptual computations follow from the nature of interactive computations, in which it becomes necessary to take into account interactions between processes during performed steps of computing (called intrastep interactions [74]). These difficulties follow also from partial information about component processes, from possible interactions between them, and also from requirements on avoidance of central control.

There are several critical issue for making progress in perception understanding and modeling. Among them is the nature of objects on which are performed computations leading to perception. We propose to use granules for modeling such objects. The computations on granules are realized through interactions.

Note also that the fusion of information may lead to new information systems with structural objects [255, 256, 254] or to nets of information systems linked by different constraints. For example, a family of parameterized sensors may model a situation in which the sensors are enabled by the judgment module for recording features of video at different moments of time in probing the environment. This makes it possible to collect the necessary features of the environment for an activating of the relevant higher level action. Parameters may be related, e.g., to positions of moving camera. This is closely related to the approach to perception presented in [138] (page 1) (see also Figure 2.8):

> *[...] perceiving is a way of acting. Perception is not something that happens to us, or in us. It is something we do. Think of blind person tap-tapping his or her way around a cluttered space, perceiving the space by touch, not all at once, but through time, by skillful probing and movement. This is, or at least ought to be, our paradigm of what perceiving is. The world makes itself available to the perceiver through physical movement and interaction.*

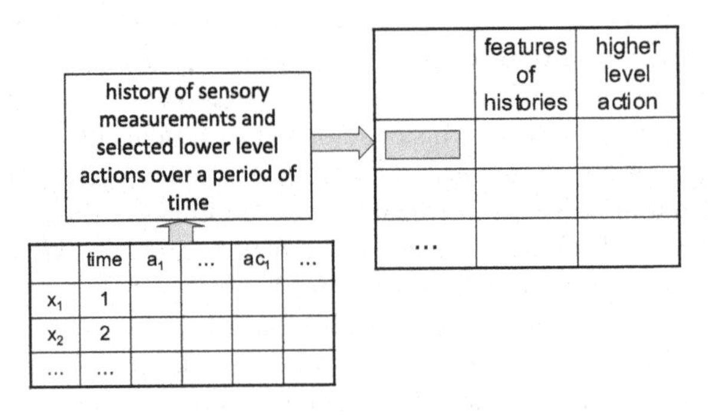

Fig. 2.8. Action in perception.

The last example suggests that the sensory attributes may be fused using some parameters such as time of enabling or position of sensors. Certainly, for performing more compound actions it is necessary to use a net of such parameterized sensors in which sensory attributes are linked by relevant constraints [138]. Hierarchical modeling may also lead to nets of information systems constructed over information systems corresponding to sensory attributes. Nodes in these networks may be linked using different information such as behavioral patterns or local theories induced from information systems in nodes as well as their changes when information systems are updated. In the former case, the reader may recognize some analogy to theory of information flow [10].

We proposed to build foundations for Perception based Computing (PBC) on the basis of Interactive Granular Computing (IGC), in particular on Interactive Rough Granular Computing (IRGC). A step toward this goal is presented in [255, 256].

PBC can be considered in a more general framework of Wisdom Technology (Wistech) [82, 84, 83] based on a metaequation

$$\textit{wisdom} = \textit{knowledge} + \textit{adaptive judgment} + \textit{interactions}. \tag{2.43}$$

In the above metaequation there is mentioned a special kind of reasoning called as *adaptive judgment*. There are many important issues belonging to adaptive judgment such as searching for relevant approximation spaces including inducing new features, feature selection, rule induction, discovery of measures of inclusion and strategies for conflict resolution, adaptation of measures based on the minimum description length, adaptive reasoning about changes, perception (action and sensory) attributes selection, adaptation of quality measures during computations performed by agents, adaptation of object structures, adaptation of strategies for knowledge representation and interaction with knowledge bases, ontology acquisition and approximation, discovery of language for cooperation or competition, and adaptive strategies for language evolution. In general, adaptive judgment is a mixture of deductive and inductive reasoning methods for reasoning about interactive granular computations and on controlling such computations by adaptive strategies for achieving the target goals. The mentioned mixture of deductive and inductive reasoning creates many challenges.

2.11 Conclusions

In the chapter, we have discussed some basic issues and methods related to rough sets together with some generalizations, including those related to relationships of rough sets with inductive reasoning. We have also listed some research directions based on interactive rough granular computing. For more detail the reader is referred to the literature cited at the beginning of this chapter (see also `http://rsds.wsiz.rzeszow.pl`).

We are observing a growing research interest in the foundations and applications of rough sets.

Relationships between rough sets and other approaches have been established as well as a wide range of hybrid systems have been developed. In particular, its relationships to fuzzy set theory, the theory of evidence, Boolean reasoning methods, statistical methods, and decision theory have been clarified and seem now to be thoroughly understood. There are reports on many hybrid methods obtained by combining the rough set approach with other approaches such as fuzzy sets, neural networks, genetic algorithms, principal component analysis, singular value decomposition or vector support machines.

Rough sets are now linked with decision system used for modeling and analysis of complex systems, fuzzy sets, neural networks, evolutionary computing, data mining and knowledge discovery, pattern recognition, machine learning, data mining, and approximate reasoning, multicriteria decision making. In particular, rough sets are used in probabilistic reasoning, granular computing, intelligent control, intelligent agent modeling, identification of autonomous systems, and process specification.

A wide range of applications of methods based on rough set theory alone or in combination with other approaches have been discovered in many areas including: acoustics, bioinformatics, business and finance, chemistry, computer engineering (e.g., data compression, digital image processing, digital signal processing, parallel and distributed computer systems, sensor fusion, fractal engineering), decision analysis and systems, economics, electrical engineering (e.g., control, signal analysis, power systems), environmental studies, digital image processing, informatics, medicine, molecular biology, musicology, neurology, robotics, social science, software engineering, spatial visualization, Web engineering, and Web mining.

Many important research topics in rough set theory such as various logics related to rough sets and many advanced algebraic properties of rough sets were only mentioned in the chapter. The reader can find details in the books, articles and journals cited in this chapter.

References

1. Aggarwal, C.: Data Streams: Models and Algorithms. Springer, Berlin (2007)
2. Alpigini, J.J., Peters, J.F., Skowron, A., Zhong, N. (eds.): RSCTC 2002. LNCS (LNAI), vol. 2475. Springer, Heidelberg (2002)
3. An, A., Stefanowski, J., Ramanna, S., Butz, C.J., Pedrycz, W., Wang, G. (eds.): RSFD-GrC 2007. LNCS (LNAI), vol. 4482. Springer, Heidelberg (2007)
4. Ariew, R., Garber, D. (eds.): Philosophical Essay, Leibniz, G. W. Hackett Publishing Company, Indianapolis (1989)
5. Balbiani, P., Vakarelov, D.: A modal logic for indiscernibility and complementarity in information systems. Fundamenta Informaticae 50(3-4), 243–263 (2002)
6. Banerjee, M., Chakraborty, M.: Logic for rough truth. Fundamenta Informaticae 71(2-3), 139–151 (2006)
7. Bargiela, A., Pedrycz, W. (eds.): Granular Computing: An Introduction. Kluwer Academic Publishers (2003)
8. Barr, B.: $*$-Autonomous categories. Lecture Notes in Mathematics, vol. 752. Springer (1979)
9. Barsalou, L.W.: Perceptual symbol systems. Behavioral and Brain Sciences 22, 577–660 (1999)
10. Barwise, J., Seligman, J.: Information Flow: The Logic of Distributed Systems. Cambridge University Press (1997)
11. Bazan, J.: Hierarchical classifiers for complex spatio-temporal concepts. In: Peters, et al. [188], pp. 474–750
12. Bazan, J.: Rough sets and granular computing in behavioral pattern identification and planning. In: Pedrycz, et al. [176], pp. 777–822
13. Bazan, J., Nguyen, H.S., Nguyen, S.H., Synak, P., Wróblewski, J.: Rough set algorithms in classification problems. In: Polkowski, et al. [200], pp. 49–88
14. Bazan, J., Skowron, A.: On-line elimination of non-relevant parts of complex objects in behavioral pattern identification. In: Pal, et al. [149], pp. 720–725
15. Bazan, J.G.: A comparison of dynamic and non-dynamic rough set methods for extracting laws from decision tables. In: Polkowski, Skowron [203], pp. 321–365
16. Bazan, J.G., Nguyen, H.S., Peters, J.F., Skowron, A., Szczuka, M., Szczuka: Rough set approach to pattern extraction from classifiers. In: Skowron, Szczuka [253], pp. 20–29, `www.elsevier.nl/locate/entcs/volume82.html`
17. Bazan, J.G., Nguyen, H.S., Skowron, A., Szczuka, M.: A view on rough set concept approximation. In: Wang, et al. [303], pp. 181–188

18. Bazan, J.G., Nguyen, H.S., Szczuka, M.S.: A view on rough set concept approximations. Fundamenta Informaticae 59, 107–118 (2004)
19. Bazan, J.G., Peters, J.F., Skowron, A.: Behavioral pattern identification through rough set modelling. In: Ślęzak, et al. [263], pp. 688–697
20. Bazan, J.G., Skowron, A.: Classifiers based on approximate reasoning schemes. In: Dunin-Kęplicz, et al. [41], pp. 191–202
21. Bazan, J.G., Skowron, A., Swiniarski, R.: Rough sets and vague concept approximation: From sample approximation to adaptive learning. In: Peters, Skowron [180], pp. 39–62
22. Behnke, S.: Hierarchical Neural Networks for Image Interpretation. LNCS, vol. 2766. Springer, Heidelberg (2003)
23. Bello, R., Falcón, R., Pedrycz, W.: Granular Computing: At the Junction of Rough Sets and Fuzzy Sets. STUDFUZZ, vol. 234. Springer, Heidelberg (2010)
24. Blake, A.: Canonical expressions in Boolean algebra. Dissertation, Dept. of Mathematics, University of Chicago, 1937. University of Chicago Libraries (1938)
25. Boole, G.: The Mathematical Analysis of Logic. G. Bell, London (1847); reprinted by Philosophical Library, New York (1948)
26. Boole, G.: An Investigation of the Laws of Thought. Walton, London (1854); reprinted by Dover Books, New York (1954)
27. Borrett, S.R., Bridewell, W., Langely, P., Arrigo, K.R.: A method for representing and developing process models. Ecological Complexity 4, 1–12 (2007)
28. Bower, J.M., Bolouri, H. (eds.): Computational Modeling of Genetic and Biochemical Networks. MIT Press (2001)
29. Breiman, L.: Statistical modeling: The two cultures. Statistical Science 16(3), 199–231 (2001)
30. Brown, F.: Boolean Reasoning. Kluwer Academic Publishers, Dordrecht (1990)
31. Cercone, N., Skowron, A., Zhong, N.: Computational Intelligence: An International Journal (Special issue) 17(3) (2001)
32. Chakraborty, M., Pagliani, P.: A Geometry of Approximation: Rough Set Theory: Logic, Algebra and Topology of Conceptual Patterns. Springer, Heidelberg (2008)
33. Chan, C.-C., Grzymala-Busse, J.W., Ziarko, W.P. (eds.): RSCTC 2008. LNCS (LNAI), vol. 5306. Springer, Heidelberg (2008)
34. Cios, K., Pedrycz, W., Swiniarski, R.: Data Mining Methods for Knowledge Discovery. Kluwer, Norwell (1998)
35. Ciucci, D., Yao, Y.Y.: Special issue on Advances in Rough Set Theory. Fundamenta Informaticae 108(3-4) (2011)
36. Delimata, P., Moshkov, M.J., Skowron, A., Suraj, Z.: Inhibitory Rules in Data Analysis: A Rough Set Approach. SCI, vol. 163. Springer, Heidelberg (2009)
37. Demri, S., Orłowska, E. (eds.): Incomplete Information: Structure, Inference, Complexity. Monographs in Theoretical Computer Sience. Springer, Heidelberg (2002)
38. Doherty, P., Łukaszewicz, W., Skowron, A., Szałas, A.: Knowledge Engineering: A Rough Set Approach. STUDFUZZ, vol. 202. Springer, Heidelberg (2006)
39. Dubois, D., Prade, H.: Foreword. In: Rough Sets: Theoretical Aspects of Reasoning about Data [169]
40. Duda, R., Hart, P., Stork, R.: Pattern Classification. John Wiley & Sons, New York (2002)
41. Dunin-Kęplicz, B., Jankowski, A., Skowron, A., Szczuka, M. (eds.): Monitoring, Security, and Rescue Tasks in Multiagent Systems (MSRAS 2004). Advances in Soft Computing. Springer, Heidelberg (2005)
42. Düntsch, I.: A logic for rough sets. Theoretical Computer Science 179, 427–436 (1997)

43. Düntsch, I., Gediga, G.: Rough set data analysis. In: Encyclopedia of Computer Science and Technology, vol. 43, pp. 281–301. Marcel Dekker (2000)
44. Düntsch, I., Gediga, G.: Rough set data analysis: A road to non-invasive knowledge discovery. Methodos Publishers, Bangor (2000)
45. Fahle, M., Poggio, T.: Perceptual Learning. MIT Press, Cambridge (2002)
46. Fan, T.F., Liau, C.J., Yao, Y.: On modal and fuzzy decision logics based on rough set theory. Fundamenta Informaticae 52(4), 323–344 (2002)
47. Feng, J., Jost, J., Minping, Q. (eds.): Network: From Biology to Theory. Springer, Berlin (2007)
48. Frege, G.: Grundgesetzen der Arithmetik, vol. 2. Verlag von Hermann Pohle, Jena (1903)
49. Friedman, J.H.: Data mining and statistics. What's the connection? (keynote address). In: Scott, D. (ed.) Proceedings of the 29th Symposium on the Interface: Computing Science and Statistics, Huston, Texas, May 14-17, University of Huston, Huston (1997)
50. Gabbay, D. (ed.): Fibring Logics. Oxford University Press (1998)
51. Gabbay, D.M., Hogger, C.J., Robinson, J.A. (eds.): Handbook of Logic in Artificial Intelligence and Logic Programming. Volume 3: Nonmonotonic Reasoning and Uncertain Reasoning. Calderon Press, Oxford (1994)
52. Garcia-Molina, H., Ullman, J., Widom, J.: Database Systems: The Complete Book. Prentice-Hall, Upper Saddle River (2002)
53. Gediga, G., Düntsch, I.: Rough approximation quality revisited. Artificial Intelligence 132, 219–234 (2001)
54. Gell-Mann, M.: The Quark and the Jaguar - Adventures in the Simple and the Complex. Brown and Co., London (1994)
55. Goldin, D., Smolka, S., Wegner, P. (eds.): Interactive Computation: The New Paradigm. Springer (2006)
56. Goldin, D., Wegner, P.: Principles of interactive computation. In: Goldin, et al. [55], pp. 25–37
57. Gomolińska, A.: A graded meaning of formulas in approximation spaces. Fundamenta Informaticae 60(1-4), 159–172 (2004)
58. Gomolińska, A.: Rough validity, confidence, and coverage of rules in approximation spaces. In: Peters, Skowron [178], pp. 57–81
59. Greco, S., Hata, Y., Hirano, S., Inuiguchi, M., Miyamoto, S., Nguyen, H.S., Słowiński, R. (eds.): RSCTC 2006. LNCS (LNAI), vol. 4259. Springer, Heidelberg (2006)
60. Greco, S., Kadzinski, M., Słowiński, R.: Selection of a representative value function in robust multiple criteria sorting. Computers & OR 38(11), 1620–1637 (2011)
61. Greco, S., Matarazzo, B., Słowiński, R.: Dealing with missing data in rough set analysis of multi-attribute and multi-criteria decision problems. In: Zanakis, S., Doukidis, G., Zopounidis, C. (eds.) Decision Making: Recent Developments and Worldwide Applications, pp. 295–316. Kluwer Academic Publishers, Boston (2000)
62. Greco, S., Matarazzo, B., Słowiński, R.: Rough set theory for multicriteria decision analysis. European Journal of Operational Research 129(1), 1–47 (2001)
63. Greco, S., Matarazzo, B., Słowiński, R.: Data mining tasks and methods: Classification: multicriteria classification. In: Kloesgen, W., Żytkow, J. (eds.) Handbook of KDD, pp. 318–328. Oxford University Press, Oxford (2002)
64. Greco, S., Matarazzo, B., Słowiński, R.: Dominance-based rough set approach to knowledge discovery (I) - General perspective (II) - Extensions and applications. In: Zhong, Liu [319], pp. 513–552, 553–612
65. Greco, S., Matarazzo, B., Słowiński, R.: Dominance-based rough set approach as a proper way of handling graduality in rough set theory. In: Peters, et al. [187], pp. 36–52

66. Greco, S., Matarazzo, B., Słowiński, R.: Granular computing and data mining for ordered data: The dominance-based rough set approach. In: Encyclopedia of Complexity and Systems Science, pp. 4283–4305. Springer, Heidelberg (2009)
67. Greco, S., Matarazzo, B., Słowiński, R.: A summary and update of "Granular computing and data mining for ordered data: The dominance-based rough set approach". In: Hu, X., Lin, T.Y., Raghavan, V.V., Grzymala-Busse, J.W., Liu, Q., Broder, A.Z. (eds.) 2010 IEEE International Conference on Granular Computing, GrC 2010, San Jose, California, August 14-16, pp. 20–21. IEEE Computer Society (2010)
68. Greco, S., Słowiński, R., Stefanowski, J., Zurawski, M.: Incremental versus non-incremental rule induction for multicriteria classification. In: Peters, et al. [183], pp. 54–62
69. Grzymała-Busse, J.W.: Managing Uncertainty in Expert Systems. Kluwer Academic Publishers, Norwell (1990)
70. Grzymała-Busse, J.W.: LERS – A system for learning from examples based on rough sets. In: Słowiński [266], pp. 3–18
71. Grzymała-Busse, J.W.: A new version of the rule induction system LERS. Fundamenta Informaticae 31(1), 27–39 (1997)
72. Grzymała-Busse, J.W.: LERS - A data mining system. In: The Data Mining and Knowledge Discovery Handbook, pp. 1347–1351 (2005)
73. Grzymala-Busse, J.W.: Generalized parameterized approximations. In: Yao, et al. [310], pp. 136–145
74. Gurevich, Y.: Interactive algorithms 2005. In: Goldin, et al. [55], pp. 165–181
75. Harnad, S.: Categorical Perception: The Groundwork of Cognition. Cambridge University Press, New York (1987)
76. Hassanien, A.E., Suraj, Z., Slezak, D., Lingras, P. (eds.): Rough Computing: Theories, Technologies and Applications. IGI Global, Hershey (2008)
77. Hastie, T., Tibshirani, R., Friedman, J.H.: The Elements of Statistical Learning: Data Mining, Inference, and Prediction. Springer, Heidelberg (2001)
78. Hirano, S., Inuiguchi, M., Tsumoto, S. (eds.): Proceedings of International Workshop on Rough Set Theory and Granular Computing (RSTGC 2001), Matsue, Shimane, Japan, May 20-22. Bulletin of the International Rough Set Society, vol. 5(1-2). International Rough Set Society, Matsue (2001)
79. Huhns, M.N., Singh, M.P.: Readings in Agents. Morgan Kaufmann, San Mateo (1998)
80. Inuiguchi, M., Hirano, S., Tsumoto, S. (eds.): Rough Set Theory and Granular Computing. STUDFUZZ, vol. 125. Springer, Heidelberg (2003)
81. Jain, R., Abraham, A.: Special issue on Hybrid Intelligence using rough sets. International Journal of Hybrid Intelligent Systems 2 (2005)
82. Jankowski, A., Skowron, A.: A wistech paradigm for intelligent systems. In: Peters, et al. [184], pp. 94–132
83. Jankowski, A., Skowron, A.: Logic for artificial intelligence: The Rasiowa - Pawlak school perspective. In: Ehrenfeucht, A., Marek, V., Srebrny, M. (eds.) Andrzej Mostowski and Foundational Studies, pp. 106–143. IOS Press, Amsterdam (2008)
84. Jankowski, A., Skowron, A.: Wisdom technology: A rough-granular approach. In: Marciniak, M., Mykowiecka, A. (eds.) Bolc Festschrift. LNCS, vol. 5070, pp. 3–41. Springer, Heidelberg (2009)
85. Jensen, R., Shen, Q.: Computational Intelligence and Feature Selection: Rough and Fuzzy Approaches. IEEE Press Series on Cmputationa Intelligence. IEEE Press and John Wiley & Sons, Hoboken, NJ (2008)
86. Jian, L., Liu, S., Lin, Y.: Hybrid Rough Sets and Applications in Uncertain Decision-Making (Systems Evaluation, Prediction, and Decision-Making). CRC Press, Boca Raton (2010)

87. Keefe, R.: Theories of Vagueness. Cambridge Studies in Philosophy, Cambridge, UK (2000)
88. Keefe, R., Smith, P.: Vagueness: A Reader. MIT Press, Massachusetts (1997)
89. Kloesgen, W., Żytkow, J. (eds.): Handbook of Knowledge Discovery and Data Mining. Oxford University Press, Oxford (2002)
90. Komorowski, J., Pawlak, Z., Polkowski, L., Skowron, A.: Rough sets: A tutorial. In: Pal, Skowron [154], pp. 3–98
91. Kostek, B.: Soft Computing in Acoustics, Applications of Neural Networks, Fuzzy Logic and Rough Sets to Physical Acoustics. STUDFUZZ, vol. 31. Physica-Verlag, Heidelberg (1999)
92. Kostek, B.: Perception-Based Data Processing in Acoustics: Applications to Music Information Retrieval and Psychophysiology of Hearing. SCI, vol. 3. Springer, Heidelberg (2005)
93. Kotlowski, W., Dembczynski, K., Greco, S., Słowiński, R.: Stochastic dominance-based rough set model for ordinal classification. Information Sciences 178(21), 4019–4037 (2008)
94. Kryszkiewicz, M., Cichoń, K.: Towards scalable algorithms for discovering rough set reducts. In: Peters, Skowron [185], pp. 120–143
95. Kryszkiewicz, M., Peters, J.F., Rybiński, H., Skowron, A. (eds.): RSEISP 2007. LNCS (LNAI), vol. 4585. Springer, Heidelberg (2007)
96. Kuznetsov, S.O., Ślęzak, D., Hepting, D.H., Mirkin, B. (eds.): RSFDGrC 2011. LNCS (LNAI), vol. 6743. Springer, Heidelberg (2011)
97. Leibniz, G.W.: Discourse on metaphysics. In: Ariew, Garber [4], pp. 35–68
98. Leśniewski, S.: Grundzüge eines neuen Systems der Grundlagen der Mathematik. Fundamenta Mathematicae 14, 1–81 (1929)
99. Leśniewski, S.: On the foundations of mathematics. Topoi 2, 7–52 (1982)
100. Lin, T.Y.: Neighborhood systems and approximation in database and knowledge base systems. In: Emrich, M.L., Phifer, M.S., Hadzikadic, M., Ras, Z.W. (eds.) Proceedings of the Fourth International Symposium on Methodologies of Intelligent Systems (Poster Session), October 12-15, pp. 75–86. Oak Ridge National Laboratory, Charlotte (1989)
101. Lin, T.Y.: Special issue, Journal of the Intelligent Automation and Soft Computing 2(2) (1996)
102. Lin, T.Y.: The discovery, analysis and representation of data dependencies in databases. In: Polkowski, L., Skowron, A. (eds.) Rough Sets in Knowledge Discovery 1: Methodology and Applications. STUDFUZZ, vol. 18, pp. 107–121. Physica-Verlag, Heidelberg (1998)
103. Lin, T.Y., Cercone, N. (eds.): Rough Sets and Data Mining - Analysis of Imperfect Data. Kluwer Academic Publishers, Boston (1997)
104. Lin, T.Y., Wildberger, A.M. (eds.): Soft Computing: Rough Sets, Fuzzy Logic, Neural Networks, Uncertainty Management, Knowledge Discovery. Simulation Councils, Inc., San Diego (1995)
105. Lin, T.Y., Yao, Y.Y., Zadeh, L.A. (eds.): Rough Sets, Granular Computing and Data Mining. STUDFUZZ. Physica-Verlag, Heidelberg (2001)
106. Liu, J.: Autonomous Agents and Multi-Agent Systems: Explorations in Learning, self-Organization and Adaptive Computation. World Scientific Publishing (2001)
107. Łukasiewicz, J.: Die logischen Grundlagen der Wahrscheinlichkeitsrechnung, 1913. In: Borkowski, L. (ed.) Jan Łukasiewicz - Selected Works, pp. 16–63. North Holland Publishing Company, Polish Scientific Publishers, Amsterdam, London, Warsaw (1970)
108. Maji, P., Pal, S.K.: Rough-Fuzzy Pattern Recognition: Application in Bioinformatics and Medical Imaging. Wiley Series in Bioinformatics. John Wiley & Sons, Hoboken (2012)

109. Marcus, S.: The paradox of the heap of grains, in respect to roughness, fuzziness and negligibility. In: Polkowski, Skowron [202], pp. 19–23
110. Marek, V.W., Rasiowa, H.: Approximating sets with equivalence relations. Theoretical Computer Science 48(3), 145–152 (1986)
111. Marek, V.W., Truszczyński, M.: Contributions to the theory of rough sets. Fundamenta Informaticae 39(4), 389–409 (1999)
112. McCarthy, J.: Notes on formalizing contex. In: Proceedings of the 13th International Joint Conference on Artifical Intelligence (IJCAI 1993), pp. 555–560. Morgan Kaufmann Publishers Inc., San Francisco (1993)
113. de Medeiros, A.K.A., Weijters, A.J.M.M., van der Aalst, W.M.P.: Genetic process mining: An experimental evaluation. Data Mining and Knowledge Discovery 14, 245–304 (2007)
114. Mill, J.S.: Ratiocinative and Inductive, Being a Connected View of the Principles of Evidence, and the Methods of Scientific Investigation. In: Parker, Son, Bourn (eds.) West Strand London (1862)
115. Mitchel, T.M.: Machine Learning. McGraw-Hill Series in Computer Science, Boston, MA (1999)
116. Moshkov, M., Skowron, A., Suraj, Z.: On testing membership to maximal consistent extensions of information systems. In: Greco, et al. [59], pp. 85–90
117. Moshkov, M., Skowron, A., Suraj, Z.: On irreducible descriptive sets of attributes for information systems. In: Chan, et al. [33], pp. 21–30
118. Moshkov, M.J., Piliszczuk, M., Zielosko, B.: Partial Covers, Reducts and Decision Rules in Rough Sets - Theory and Applications. SCI, vol. 145. Springer, Heidelberg (2008)
119. Moshkov, M.J., Skowron, A., Suraj, Z.: On minimal rule sets for almost all binary information systems. Fundamenta Informaticae 80(1-3), 247–258 (2007)
120. Moshkov, M.J., Skowron, A., Suraj, Z.: On minimal inhibitory rules for almost all k-valued information systems. Fundamenta Informaticae 93(1-3), 261–272 (2009)
121. Moshkov, M., Zielosko, B.: Combinatorial Machine Learning - A Rough Set Approach. SCI, vol. 360. Springer, Heidelberg (2011)
122. Nakamura, A.: Fuzzy quantifiers and rough quantifiers. In: Wang, P.P. (ed.) Advances in Fuzzy Theory and Technology II, pp. 111–131. Duke University Press, Durham (1994)
123. Nakamura, A.: On a logic of information for reasoning about knowledge. In: Ziarko [322], pp. 186–195
124. Nakamura, A.: A rough logic based on incomplete information and its application. International Journal of Approximate Reasoning 15(4), 367–378 (1996)
125. Nguyen, H.S.: Efficient SQL-learning method for data mining in large data bases. In: Dean, T. (ed.) Sixteenth International Joint Conference on Artificial Intelligence, IJCAI, pp. 806–811. Morgan-Kaufmann Publishers, Stockholm (1999)
126. Nguyen, H.S.: On efficient handling of continuous attributes in large data bases. Fundamenta Informaticae 48(1), 61–81 (2001)
127. Nguyen, H.S.: Approximate boolean reasoning approach to rough sets and data mining. In: Ślęzak, et al. [263], pp. 12–22 (plenary talk)
128. Nguyen, H.S.: Approximate boolean reasoning: Foundations and applications in data mining. In: Peters, Skowron [180], pp. 344–523
129. Nguyen, H.S., Jankowski, A., Skowron, A., Stepaniuk, J., Szczuka, M.: Discovery of process models from data and domain knowledge: A rough-granular approach. In: Yao, J.T. (ed.) Novel Developments in Granular Computing: Applications for Advanced Human Reasoning and Soft Computation, pp. 16–47. IGI Global, Hershey (2010)

130. Nguyen, H.S., Nguyen, S.H.: Rough sets and association rule generation. Fundamenta Informaticae 40(4), 383–405 (1999)
131. Nguyen, H.S., Skowron, A.: A rough granular computing in discovery of process models from data and domain knowledge. Journal of Chongqing University of Post and Telecommunications 20(3), 341–347 (2008)
132. Nguyen, H.S., Ślęzak, D.: Approximate reducts and association rules - correspondence and complexity results. In: Skowron, et al. [234], pp. 137–145
133. Nguyen, S.H., Bazan, J., Skowron, A., Nguyen, H.S.: Layered learning for concept synthesis. In: Peters, Skowron [185], pp. 187–208
134. Nguyen, S.H., Nguyen, H.S.: Some efficient algorithms for rough set methods. In: Sixth International Conference on Information Processing and Management of Uncertainty on Knowledge Based Systems, IPMU 1996, Granada, Spain, vol. III, pp. 1451–1456 (1996)
135. Nguyen, T.T.: Eliciting domain knowledge in handwritten digit recognition. In: Pal, et al. [149], pp. 762–767
136. Nguyen, T.T., Skowron, A.: Rough set approach to domain knowledge approximation. In: Wang, et al. [303], pp. 221–228
137. Nguyen, T.T., Skowron, A.: Rough-granular computing in human-centric information processing. In: Bargiela, A., Pedrycz, W. (eds.) Human-Centric Information Processing Through Granular Modelling. SCI, vol. 182, pp. 1–30. Springer, Heidelberg (2009)
138. Noë, A.: Action in Perception. MIT Press (2004)
139. Omicini, A., Ricci, A., Viroli, M.: The multidisciplinary patterns of interaction from sciences to computer science. In: Goldin, et al. [55], pp. 395–414
140. Orłowska, E.: Semantics of vague concepts. In: Dorn, G., Weingartner, P. (eds.) Foundation of Logic and Linguistics, pp. 465–482. Plenum Press, New York (1984)
141. Orłowska, E.: Rough concept logic. In: Skowron [224], pp. 177–186
142. Orłowska, E.: Reasoning about vague concepts. Bulletin of the Polish Academy of Sciences, Mathematics 35, 643–652 (1987)
143. Orłowska, E.: Logic for reasoning about knowledge. Zeitschrift für Mathematische Logik und Grundlagen der Mathematik 35, 559–572 (1989)
144. Orłowska, E.: Kripke semantics for knowledge representation logics. Studia Logica 49(2), 255–272 (1990)
145. Orłowska, E. (ed.): Incomplete Information: Rough Set Analysis. STUDFUZZ, vol. 13. Springer/Physica-Verlag, Heidelberg (1997)
146. Orłowska, E., Pawlak, Z.: Representation of non-deterministic information. Theoretical Computer Science 29, 27–39 (1984)
147. Orłowska, E., Peters, J.F., Rozenberg, G., Skowron, A.: Special volume dedicated to the memory of Zdzisław Pawlak. Fundamenta Informaticae 75(1-4) (2007)
148. Pal, S.: Computational theory perception (CTP), rough-fuzzy uncertainty analysis and mining in bioinformatics and web intelligence: A unified framework. In: Peters, Skowron [182], pp. 106–129
149. Pal, S.K., Bandyopadhyay, S., Biswas, S. (eds.): PReMI 2005. LNCS, vol. 3776. Springer, Heidelberg (2005)
150. Pal, S.K., Mitra, P.: Pattern Recognition Algorithms for Data Mining. CRC Press, Boca Raton (2004)
151. Pal, S.K., Pedrycz, W., Skowron, A., Swiniarski, R.: Special volume: Rough-neuro computing. Neurocomputing 36 (2001)
152. Pal, S.K., Peters, J.F. (eds.): Rough Fuzzy Image Analysis Foundations and Methodologies. Chapman & Hall/CRC, Boca Raton, Fl (2010)

153. Pal, S.K., Polkowski, L., Skowron, A. (eds.): Rough-Neural Computing: Techniques for Computing with Words. Cognitive Technologies. Springer, Heidelberg (2004)
154. Pal, S.K., Skowron, A. (eds.): Rough Fuzzy Hybridization: A New Trend in Decision-Making. Springer, Singapore (1999)
155. Pancerz, K., Suraj, Z.: Modelling concurrent systems specified by dynamic information systems: A rough set approach. Electronic Notes in Theoretical Computer Science 82(4), 206–218 (2003)
156. Pancerz, K., Suraj, Z.: Discovering concurrent models from data tables with the ROSECON system. Fundamenta Informaticae 60(1-4), 251–268 (2004)
157. Pancerz, K., Suraj, Z.: Discovering concurrent models from data tables with the ROSECON system. Fundamenta Informaticae 60(1-4), 251–268 (2004)
158. Pancerz, K., Suraj, Z.: Discovery of asynchronous concurrent models from experimental tables. Fundamenta Informaticae 61(2), 97–116 (2004)
159. Pancerz, K., Suraj, Z.: Restriction-based concurrent system design using the rough set formalism. Fundamenta Informaticae 67(1-3), 233–247 (2005)
160. Pancerz, K., Suraj, Z.: Reconstruction of concurrent system models described by decomposed data tables. Fundamenta Informaticae 71(1), 121–137 (2006)
161. Pancerz, K., Suraj, Z.: Towards efficient computing consistent and partially consistent extensions of information systems. Fundamenta Informaticae 79(3-4), 553–566 (2007)
162. Papageorgiou, E.I., Stylios, C.D.: Fuzzy cognitive maps. In: Pedrycz, et al. [176], pp. 755–774
163. Pawlak, Z.: Classification of Objects by Means of Attributes, Reports, vol. 429, Institute of Computer Science, Polish Academy of Sciences, Warsaw, Poland (1981)
164. Pawlak, Z.: Information systems - theoretical foundations. Information Systems 6, 205–218 (1981)
165. Pawlak, Z.: Rough Relations, Reports, vol. 435. Institute of Computer Science, Polish Academy of Sciences, Warsaw, Poland (1981)
166. Pawlak, Z.: Rough sets. International Journal of Computer and Information Sciences 11, 341–356 (1982)
167. Pawlak, Z.: Rough logic. Bulletin of the Polish Academy of Sciences, Technical Sciences 35(5-6), 253–258 (1987)
168. Pawlak, Z.: Decision logic. Bulletin of the EATCS 44, 201–225 (1991)
169. Pawlak, Z.: Rough Sets: Theoretical Aspects of Reasoning about Data. In: System Theory, Knowledge Engineering and Problem Solving, vol. 9. Kluwer Academic Publishers, Dordrecht (1991)
170. Pawlak, Z.: Concurrent versus sequential - the rough sets perspective. Bulletin of the EATCS 48, 178–190 (1992)
171. Pawlak, Z.: Decision rules, Bayes' rule and rough sets. In: Skowron, et al. [234], pp. 1–9
172. Pawlak, Z., Skowron, A.: Rough membership functions. In: Yager, R., Fedrizzi, M., Kacprzyk, J. (eds.) Advances in the Dempster-Shafer Theory of Evidence, pp. 251–271. John Wiley & Sons, New York (1994)
173. Pawlak, Z., Skowron, A.: Rough sets and boolean reasoning. Information Sciences 177(1), 41–73 (2007)
174. Pawlak, Z., Skowron, A.: Rough sets: Some extensions. Information Sciences 177(28-40), 1 (2007)
175. Pawlak, Z., Skowron, A.: Rudiments of rough sets. Information Sciences 177(1), 3–27 (2007)
176. Pedrycz, W., Skowron, S., Kreinovich, V. (eds.): Handbook of Granular Computing. John Wiley & Sons, Hoboken (2008)

177. Peters, J., Skowron, A.: Special issue on a rough set approach to reasoning about data. International Journal of Intelligent Systems 16(1) (2001)
178. Peters, J.F., Skowron, A. (eds.): Transactions on Rough Sets III. LNCS, vol. 3400. Springer, Heidelberg (2005)
179. Peters, J.F., Skowron, A. (eds.): Transactions on Rough Sets IV. LNCS, vol. 3700. Springer, Heidelberg (2005)
180. Peters, J.F., Skowron, A. (eds.): Transactions on Rough Sets V. LNCS, vol. 4100. Springer, Heidelberg (2006)
181. Peters, J.F., Skowron, A. (eds.): Transactions on Rough Sets VIII. LNCS, vol. 5084. Springer, Heidelberg (2008)
182. Peters, J.F., Skowron, A. (eds.): Transactions on Rough Sets XI. LNCS, vol. 5946. Springer, Heidelberg (2010)
183. Peters, J.F., Skowron, A., Dubois, D., Grzymała-Busse, J.W., Inuiguchi, M., Polkowski, L. (eds.): Transactions on Rough Sets II. LNCS, vol. 3135. Springer, Heidelberg (2004)
184. Peters, J.F., Skowron, A., Düntsch, I., Grzymała-Busse, J.W., Orłowska, E., Polkowski, L. (eds.): Transactions on Rough Sets VI. LNCS, vol. 4374. Springer, Heidelberg (2007)
185. Peters, J.F., Skowron, A., Grzymała-Busse, J.W., Kostek, B.z., Świniarski, R.W., Szczuka, M.S. (eds.): Transactions on Rough Sets I. LNCS, vol. 3100. Springer, Heidelberg (2004)
186. Peters, J.F., Skowron, A., Chan, C.-C., Grzymala-Busse, J.W., Ziarko, W.P. (eds.): Transactions on Rough Sets XIII. LNCS, vol. 6499. Springer, Heidelberg (2011)
187. Peters, J.F., Skowron, A., Marek, V.W., Orłowska, E., Słowiński, R., Ziarko, W.P. (eds.): Transactions on Rough Sets VII. LNCS, vol. 4400. Springer, Heidelberg (2007)
188. Peters, J.F., Skowron, A., Rybiński, H. (eds.): Transactions on Rough Sets IX. LNCS, vol. 5390. Springer, Heidelberg (2008)
189. Peters, J.F., Skowron, A., Sakai, H., Chakraborty, M.K., Ślęzak, D., Hassanien, A.E., Zhu, W.: Transactions on Rough Sets XIV. LNCS, vol. 6600. Springer, Heidelberg (2011)
190. Peters, J.F., Skowron, A., Słowiński, R., Lingras, P., Miao, D., Tsumoto, S. (eds.): Transactions on Rough Sets XII. LNCS, vol. 6190. Springer, Heidelberg (2010)
191. Peters, J.F., Skowron, A., Suraj, Z.: An application of rough set methods in control design. Fundamenta Informaticae 43(1-4), 269–290 (2000)
192. Peters, J.F., Skowron, A., Suraj, Z.: An application of rough set methods in control design. Fundamenta Informaticae 43(1-4), 269–290 (2000)
193. Peters, J.F., Skowron, A., Wolski, M., Chakraborty, M.K., Wu, W.-Z. (eds.): Transactions on Rough Sets X. LNCS, vol. 5656. Springer, Heidelberg (2009)
194. Pindur, R., Susmaga, R., Stefanowski, J.: Hyperplane aggregation of dominance decision rules. Fundamenta Informaticae 61(2), 117–137 (2004)
195. Poggio, T., Smale, S.: The mathematics of learning: Dealing with data. Notices of the AMS 50(5), 537–544 (2003)
196. Polkowski, L.: Rough Sets: Mathematical Foundations. Advances in Soft Computing. Physica-Verlag, Heidelberg (2002)
197. Polkowski, L.: Rough mereology: A rough set paradigm for unifying rough set theory and fuzzy set theory. Fundamenta Informaticae 54, 67–88 (2003)
198. Polkowski, L.: A note on 3-valued rough logic accepting decision rules. Fundamenta Informaticae 61(1), 37–45 (2004)
199. Polkowski, L.: Approximate Reasoning by Parts. An Introduction to Rough Mereology. ISRL, vol. 20. Springer, Heidelberg (2011)

200. Polkowski, L., Lin, T.Y., Tsumoto, S. (eds.): Rough Set Methods and Applications: New Developments in Knowledge Discovery in Information Systems. STUDFUZZ, vol. 56. Springer/Physica-Verlag, Heidelberg (2000)
201. Polkowski, L., Skowron, A.: Rough mereology: A new paradigm for approximate reasoning. International Journal of Approximate Reasoning 15(4), 333–365 (1996)
202. Polkowski, L., Skowron, A. (eds.): RSCTC 1998. LNCS (LNAI), vol. 1424. Springer, Heidelberg (1998)
203. Polkowski, L., Skowron, A. (eds.): Rough Sets in Knowledge Discovery 1: Methodology and Applications. STUDFUZZ, vol. 18. Physica-Verlag, Heidelberg (1998)
204. Polkowski, L., Skowron, A. (eds.): Rough Sets in Knowledge Discovery 2: Applications, Case Studies and Software Systems. STUDFUZZ, vol. 19. Physica-Verlag, Heidelberg (1998)
205. Polkowski, L., Skowron, A.: Towards adaptive calculus of granules. In: Zadeh, L.A., Kacprzyk, J. (eds.) Computing with Words in Information/Intelligent Systems, pp. 201–227. Physica-Verlag, Heidelberg (1999)
206. Polkowski, L., Skowron, A.: Rough mereological calculi of granules: A rough set approach to computation. Computational Intelligence: An International Journal 17(3), 472–492 (2001)
207. Polkowski, L., Skowron, A., Żytkow, J.: Rough foundations for rough sets. In: Lin, Wildberger [104], pp. 55–58
208. Ramsay, J.O., Silverman, B.W.: Applied Functional Data Analysis. Springer, Berlin (2002)
209. Rasiowa, H.: Axiomatization and completeness of uncountably valued approximation logic. Studia Logica 53(1), 137–160 (1994)
210. Rasiowa, H., Skowron, A.: Approximation logic. In: Bibel, W., Jantke, K.P. (eds.) Mathematical Methods of Specification and Synthesis of Software Systems. Mathematical Research, vol. 31, pp. 123–139. Akademie Verlag, Berlin (1985)
211. Rasiowa, H., Skowron, A.: Rough concept logic. In: Skowron [224], pp. 288–297
212. Rauszer, C.: An equivalence between indiscernibility relations in information systems and a fragment of intuitionistic logic. In: Skowron [224], pp. 298–317
213. Rauszer, C.: An equivalence between theory of functional dependence and a fragment of intuitionistic logic. Bulletin of the Polish Academy of Sciences, Mathematics 33, 571–579 (1985)
214. Rauszer, C.: Logic for information systems. Fundamenta Informaticae 16, 371–382 (1992)
215. Rauszer, C.: Knowledge representation systems for groups of agents. In: Wroński, J. (ed.) Philosophical Logic in Poland, pp. 217–238. Kluwer, Dordrecht (1994)
216. Read, S.: Thinking about Logic: An Introduction to the Philosophy of Logic. Oxford University Press, Oxford (1994)
217. Rissanen, J.: Modeling by shortes data description. Automatica 14, 465–471 (1978)
218. Rissanen, J.: Minimum-description-length principle. In: Kotz, S., Johnson, N. (eds.) Encyclopedia of Statistical Sciences, pp. 523–527. John Wiley & Sons, New York (1985)
219. Roddick, J., Hornsby, K.S., Spiliopoulou, M.: An Updated Bibliography of Temporal, Spatial, and Spatio-temporal Data Mining Research. In: Roddick, J., Hornsby, K.S. (eds.) TSDM 2000. LNCS (LNAI), vol. 2007, pp. 147–164. Springer, Heidelberg (2001)
220. Russell, B.: An Inquiry into Meaning and Truth. George Allen & Unwin Ltd. and W. W. Norton, London and New York (1940)
221. Sakai, H., Chakraborty, M.K., Hassanien, A.E., Ślęzak, D., Zhu, W. (eds.): RSFDGrC 2009. LNCS, vol. 5908. Springer, Heidelberg (2009)

222. Serafini, L., Bouquet, P.: Comparing formal theories of context in ai. Artificial Intelligence 155, 41–67 (2004)
223. Skowron, A.: Rough Sets in Perception-Based Computing (keynote talk). In: Pal, S.K., Bandyopadhyay, S., Biswas, S. (eds.) PReMI 2005. LNCS, vol. 3776, pp. 21–29. Springer, Heidelberg (2005)
224. Skowron, A. (ed.): SCT 1984. LNCS, vol. 208. Springer, Heidelberg (1985)
225. Skowron, A.: Boolean Reasoning for Decision Rules Generation. In: Komorowski, J., Raś, Z.W. (eds.) ISMIS 1993. LNCS, vol. 689, pp. 295–305. Springer, Heidelberg (1993)
226. Skowron, A.: Extracting laws from decision tables. Computational Intelligence: An International Journal 11, 371–388 (1995)
227. Skowron, A.: Rough sets in KDD - plenary talk. In: Shi, Z., Faltings, B., Musen, M. (eds.) 16th World Computer Congress (IFIP 2000): Proceedings of Conference on Intelligent Information Processing (IIP 2000), pp. 1–14. Publishing House of Electronic Industry, Beijing (2000)
228. Skowron, A.: Approximate reasoning by agents in distributed environments. In: Zhong, N., Liu, J., Ohsuga, S., Bradshaw, J. (eds.) Intelligent Agent Technology Research and Development: Proceedings of the 2nd Asia-Pacific Conference on Intelligent Agent Technology, IAT 2001, Maebashi, Japan, October 23-26, pp. 28–39. World Scientific, Singapore (2001)
229. Skowron, A.: Toward intelligent systems: Calculi of information granules. Bulletin of the International Rough Set Society 5(1-2), 9–30 (2001)
230. Skowron, A.: Approximate reasoning in distributed environments. In: Zhong, Liu [319], pp. 433–474
231. Skowron, A.: Perception logic in intelligent systems (keynote talk). In: Blair, S., et al. (eds.) Proceedings of the 8th Joint Conference on Information Sciences (JCIS 2005), Salt Lake City, Utah, July 21-26, pp. 1–5. X-CD Technologies: A Conference & Management Company, Toronto (2005)
232. Skowron, A.: Rough sets and vague concepts. Fundamenta Informaticae 64(1-4), 417–431 (2005)
233. Skowron, A., Grzymała-Busse, J.W.: From rough set theory to evidence theory. In: Yager, R., Fedrizzi, M., Kacprzyk, J. (eds.) Advances in the Dempster-Shafer Theory of Evidence, pp. 193–236. John Wiley & Sons, New York (1994)
234. Skowron, A., Ohsuga, S., Zhong, N. (eds.): RSFDGrC 1999. LNCS (LNAI), vol. 1711. Springer, Heidelberg (1999)
235. Skowron, A., Pal, S.K.: Special volume: Rough sets, pattern recognition and data mining. Pattern Recognition Letters 24(6) (2003)
236. Skowron, A., Pal, S.K., Nguyen, H.S.: Special issue: Rough sets and fuzzy sets in natural computing. Theoretical Computer Science 412(42) (2011)
237. Skowron, A., Peters, J.: Rough sets: Trends and challenges. In: Wang, et al. [303], pp. 25–34 (plenary talk)
238. Skowron, A., Rauszer, C.: The discernibility matrices and functions in information systems. In: Słowiński [266], pp. 331–362
239. Skowron, A., Stepaniuk, J.: Generalized approximation spaces. In: The Third International Workshop on Rough Sets and Soft Computing Proceedings (RSSC 1994), San Jose, California, USA, November 10-12, pp. 156–163 (1994)
240. Skowron, A., Stepaniuk, J.: Tolerance approximation spaces. Fundamenta Informaticae 27(2-3), 245–253 (1996)
241. Skowron, A., Stepaniuk, J.: Information granules: Towards foundations of granular computing. International Journal of Intelligent Systems 16(1), 57–86 (2001)

242. Skowron, A., Stepaniuk, J.: Information granules and rough-neural computing. In: Pal, et al. [153], pp. 43–84
243. Skowron, A., Stepaniuk, J.: Ontological framework for approximation. In: Ślęzak, et al. [262], pp. 718–727
244. Skowron, A., Stepaniuk, J.: Approximation spaces in rough-granular computing. Fundamenta Informaticae 100, 141–157 (2010)
245. Skowron, A., Stepaniuk, J., Peters, J., Swiniarski, R.: Calculi of approximation spaces. Fundamenta Informaticae 72, 363–378 (2006)
246. Skowron, A., Stepaniuk, J., Swiniarski, R.: Modeling rough granular computing based on approximation spaces. Information Sciences 184, 20–43 (2012)
247. Skowron, A., Suraj, Z.: A rough set approach to real-time state identification. Bulletin of the EATCS 50, 264–275 (1993)
248. Skowron, A., Suraj, Z.: Rough sets and concurrency. Bulletin of the Polish Academy of Sciences, Technical Sciences 41, 237–254 (1993)
249. Skowron, A., Suraj, Z.: Discovery of concurrent data models from experimental tables: A rough set approach. In: Proceedings of the First International Conference on Knowledge Discovery and Data Mining (KDD 1995), Montreal, Canada, August 20-21, pp. 288–293. AAAI Press, Menlo Park (1995)
250. Skowron, A., Swiniarski, R.: Rough sets and higher order vagueness. In: Ślęzak, et al. [262], pp. 33–42
251. Skowron, A., Swiniarski, R., Synak, P.: Approximation spaces and information granulation. In: Peters, Skowron [178], pp. 175–189
252. Skowron, A., Synak, P.: Complex patterns. Fundamenta Informaticae 60(1-4), 351–366 (2004)
253. Skowron, A., Szczuka, M. (eds.): Proceedings of the Workshop on Rough Sets in Knowledge Discovery and Soft Computing at ETAPS 2003, April 12-13. Electronic Notes in Computer Science, vol. 82(4). Elsevier, Amsterdam (2003), `www.elsevier.nl/locate/entcs/volume82.html`
254. Skowron, A., Szczuka, M.: Toward Interactive Computations: A Rough-Granular Approach. In: Koronacki, J., Raś, Z.W., Wierzchoń, S.T., Kacprzyk, J. (eds.) Advances in Machine Learning II. SCI, vol. 263, pp. 23–42. Springer, Heidelberg (2010)
255. Skowron, A., Wasilewski, P.: Information systems in modeling interactive computations on granules. Theoretical Computer Science 412(42), 5939–5959 (2011)
256. Skowron, A., Wasilewski, P.: Toward interactive rough-granular computing. Control & Cybernetics 40(2), 1–23 (2011)
257. Ślęzak, D.: Approximate reducts in decision tables. In: Sixth International Conference on Information Processing and Management of Uncertainty in Knowledge-Based Systems, IPMU 1996, vol. III, pp. 1159–1164. Granada, Spain (1996)
258. Ślęzak, D.: Normalized decision functions and measures for inconsistent decision tables analysis. Fundamenta Informaticae 44, 291–319 (2000)
259. Ślęzak, D.: Various approaches to reasoning with frequency-based decision reducts: A survey. In: Polkowski, et al. [200], pp. 235–285
260. Ślęzak, D.: Approximate entropy reducts. Fundamenta Informaticae 53, 365–387 (2002)
261. Ślęzak, D.: Rough sets and Bayes factor. In: Peters, Skowron [178], pp. 202–229
262. Ślęzak, D., Wang, G., Szczuka, M.S., Düntsch, I., Yao, Y. (eds.): RSFDGrC 2005, Part I. LNCS (LNAI), vol. 3641. Springer, Heidelberg (2005)
263. Ślęzak, D., Yao, J., Peters, J.F., Ziarko, W.P., Hu, X. (eds.): RSFDGrC 2005, Part II. LNCS (LNAI), vol. 3642. Springer, Heidelberg (2005)

264. Ślęzak, D., Ziarko, W.: The investigation of the Bayesian rough set model. International Journal of Approximate Reasoning 40, 81–91 (2005)
265. Słowiński, R.: New Applications and Theoretical Foundations of the Dominance-based Rough Set Approach. In: Szczuka, M., Kryszkiewicz, M., Ramanna, S., Jensen, R., Hu, Q. (eds.) RSCTC 2010. LNCS, vol. 6086, pp. 2–3. Springer, Heidelberg (2010)
266. Słowiński, R. (ed.): Intelligent Decision Support - Handbook of Applications and Advances of the Rough Sets Theory. System Theory, Knowledge Engineering and Problem Solving, vol. 11. Kluwer Academic Publishers, Dordrecht (1992)
267. Słowiński, R., Greco, S., Matarazzo, B.: Rough set analysis of preference-ordered data. In: Alpigini, et al. [2], pp. 44–59
268. Słowiński, R., Stefanowski, J. (eds.): Special issue: Proceedings of the First International Workshop on Rough Sets: State of the Art and Perspectives, Kiekrz, Poznań, Poland, September 2-4 (1992); Foundations of Computing and Decision Sciences 18(3-4) (1993)
269. Sowa, J.F.: Knowledge Representation: Logical, Philosophical, and Computational Foundations. Brooks Cole Publishing Co. (2000)
270. Staab, S., Studer, R. (eds.): Handbook on Ontologies. International Handbooks on Information Systems. Springer, Heidelberg (2004)
271. Stepaniuk, J.: Approximation spaces, reducts and representatives. In: Polkowski, Skowron [204], pp. 109–126
272. Stepaniuk, J. (ed.): Rough-Granular Computing in Knowledge Discovery and Data Mining. Springer, Heidelberg (2008)
273. Stone, P.: Layered Learning in Multi-Agent Systems: A Winning Approach to Robotic Soccer. The MIT Press, Cambridge (2000)
274. Strąkowski, T., Rybiński, H.: A new approach to distributed algorithms for reduct calculation. In: Peters, Skowron [188], pp. 365–378
275. Suraj, Z.: Discovery of concurrent data models from experimental tables: A rough set approach. Fundamenta Informaticae 28(3-4), 353–376 (1996)
276. Suraj, Z.: Rough set methods for the synthesis and analysis of concurrent processes. In: Polkowski, et al. [200], pp. 379–488
277. Suraj, Z.: Discovering concurrent process models in data: A rough set approach. In: Sakai, et al. [221], pp. 12–19
278. Suraj, Z., Pancerz, K.: A synthesis of concurrent systems: A rough set approach. In: Wang, et al. [303], pp. 299–302
279. Suraj, Z., Pancerz, K.: The rosecon system - a computer tool for modelling and analysing of processes. In: 2005 International Conference on Computational Intelligence for Modelling Control and Automation (CIMCA 2005), International Conference on Intelligent Agents, Web Technologies and Internet Commerce (IAWTIC 2005), Vienna, Austria, November 28-30, pp. 829–834. IEEE Computer Society (2005)
280. Suraj, Z., Pancerz, K.: Some remarks on computing consistent extensions of dynamic information systems. In: Proceedings of the Fifth International Conference on Intelligent Systems Design and Applications (ISDA 2005), Wrocław, Poland, September 8-10, pp. 420–425. IEEE Computer Society (2005)
281. Suraj, Z., Pancerz, K., Owsiany, G.: On consistent and partially consistent extensions of information systems. In: Ślęzak et al. [262], pp. 224–233
282. Swift, J.: Gulliver's Travels into Several Remote Nations of the World (ananymous publisher), London, M, DCC, XXVI (1726)
283. Swiniarski, R.W., Pancerz, K., Suraj, Z.: Prediction of model changes of concurrent systems described by temporal information systems. In: Proceedings of The 2005 International Conference on Data Mining (DMIN 2005), Las Vegas, Nevada, USA, June 20-23, pp. 51–57. CSREA Press (2005)

284. Sycara, K.: Multiagent systems. AI Magazine, 79–92 (Summer 1998)
285. Szczuka, M., Skowron, A., Stepaniuk, J.: Function approximation and quality measures in rough-granular systems. Fundamenta Informaticae 109(3-4), 339–354 (2011)
286. Szczuka, M.S., Kryszkiewicz, M., Ramanna, S., Jensen, R., Hu, Q. (eds.): RSCTC 2010. LNCS, vol. 6086. Springer, Heidelberg (2010)
287. Tarski, A.: Logic, Semantics, Metamathematics. Oxford University Press, Oxford (1983) [translated by J. H. Woodger]
288. Taylor, G.W., Fergus, R., LeCun, Y., Bregler, C.: Convolutional Learning of Spatio-Temporal Features. In: Daniilidis, K., Maragos, P., Paragios, N. (eds.) ECCV 2010. LNCS, vol. 6316, pp. 140–153. Springer, Heidelberg (2010)
289. Terano, T., Nishida, T., Namatame, A., Tsumoto, S., Ohsawa, Y., Washio, T. (eds.): JSAI-WS 2001. LNCS (LNAI), vol. 2253. Springer, Heidelberg (2001)
290. Torra, V., Narukawa, Y.: Modeling Decisions Information Fusion and Aggregation Operators. Springer (2007)
291. Tsumoto, S., Kobayashi, S., Yokomori, T., Tanaka, H., Nakamura, A. (eds.): Proceedings of the The Fourth International Workshop on Rough Sets, Fuzzy Sets and Machine Discovery, November 6-8. University of Tokyo, Tokyo (1996)
292. Tsumoto, S., Słowiński, R., Komorowski, J., Grzymała-Busse, J.W. (eds.): RSCTC 2004. LNCS (LNAI), vol. 3066. Springer, Heidelberg (2004)
293. Tsumoto, S., Tanaka, H.: PRIMEROSE: Probabilistic rule induction method based on rough sets and resampling methods. Computational Intelligence: An International Journal 11, 389–405 (1995)
294. Unnikrishnan, K.P., Ramakrishnan, N., Sastry, P.S., Uthurusamy, R.: Network reconstruction from dynamic data. SIGKDD Explorations 8(2), 90–91 (2006)
295. Vakarelov, D.: A modal logic for similarity relations in Pawlak knowledge representation systems. Fundamenta Informaticae 15(1), 61–79 (1991)
296. Vakarelov, D.: Modal logics for knowledge representation systems. Theoretical Computer Science 90(2), 433–456 (1991)
297. Vakarelov, D.: A duality between Pawlak's knowledge representation systems and bi-consequence systems. Studia Logica 55(1), 205–228 (1995)
298. Vakarelov, D.: A modal characterization of indiscernibility and similarity relations in Pawlak's information systems. In: Ślęzak et al. [262], pp. 12–22 (plenary talk)
299. van der Aalst, W.M.P. (ed.): Process Mining Discovery, Conformance and Enhancement of Business Processes. Springer, Heidelberg (2011)
300. Vapnik, V.: Statistical Learning Theory. John Wiley & Sons, New York (1998)
301. Vitória, A.: A framework for reasoning with rough sets. Licentiate Thesis, Linköping University 2004. In: Peters, Skowron [179], pp. 178–276
302. Wang, G., Li, T.R., Grzymala-Busse, J.W., Miao, D., Skowron, A., Yao, Y. (eds.): RSKT 2008. LNCS (LNAI), vol. 5009. Springer, Heidelberg (2008)
303. Wang, G., Liu, Q., Yao, Y., Skowron, A. (eds.): RSFDGrC 2003. LNCS (LNAI), vol. 2639. Springer, Heidelberg (2003)
304. Wang, G.-Y., Peters, J.F., Skowron, A., Yao, Y. (eds.): RSKT 2006. LNCS (LNAI), vol. 4062. Springer, Heidelberg (2006)
305. Wen, P., Li, Y., Polkowski, L., Yao, Y., Tsumoto, S., Wang, G. (eds.): RSKT 2009. LNCS, vol. 5589. Springer, Heidelberg (2009)
306. Wong, S.K.M., Ziarko, W.: Comparison of the probabilistic approximate classification and the fuzzy model. Fuzzy Sets and Systems 21, 357–362 (1987)
307. Wróblewski, J.: Theoretical foundations of order-based genetic algorithms. Fundamenta Informaticae 28, 423–430 (1996)

308. Wu, F.X.: Inference of gene regulatory networks and its validation. Current Bioinformatics 2(2), 139–144 (2007)
309. Yao, J., Lingras, P., Wu, W.-Z., Szczuka, M.S., Cercone, N.J., Ślęzak, D. (eds.): RSKT 2007. LNCS (LNAI), vol. 4481. Springer, Heidelberg (2007)
310. Yao, J., Ramanna, S., Wang, G., Suraj, Z. (eds.): RSKT 2011. LNCS, vol. 6954. Springer, Heidelberg (2011)
311. Yao, Y.Y.: Generalized rough set models. In: Polkowski, Skowron [203], pp. 286–318
312. Yao, Y.Y.: Information granulation and rough set approximation. International Journal of Intelligent Systems 16, 87–104 (2001)
313. Yao, Y.Y.: On generalizing rough set theory. In: Wang, et al. [303], pp. 44–51
314. Yao, Y.Y.: Probabilistic approaches to rough sets. Expert Systems 20, 287–297 (2003)
315. Yao, Y.Y., Wong, S.K.M., Lin, T.Y.: A review of rough set models. In: Lin, Cercone [103], pp. 47–75
316. Yu, J., Greco, S., Lingras, P., Wang, G., Skowron, A. (eds.): RSKT 2010. LNCS, vol. 6401. Springer, Heidelberg (2010)
317. Zadeh, L.A.: Fuzzy sets. Information and Control 8, 338–353 (1965)
318. Zadeh, L.A.: A new direction in AI: Toward a computational theory of perceptions. AI Magazine 22(1), 73–84 (2001)
319. Zhong, N., Liu, J. (eds.): Intelligent Technologies for Information Analysis. Springer, Heidelberg (2004)
320. Zhu, W.: Topological approaches to covering rough sets. Information Sciences 177, 1499–1508 (2007)
321. Ziarko, W.: Variable precision rough set model. Journal of Computer and System Sciences 46, 39–59 (1993)
322. Ziarko, W. (ed.): Rough Sets, Fuzzy Sets and Knowledge Discovery: Proceedings of the Second International Workshop on Rough Sets and Knowledge Discovery (RSKD 1993), Banff, Alberta, Canada, October 12-15 (1993); Workshops in Computing. Springer & British Computer Society, London, Berlin (1994)
323. Ziarko, W.: Special issue, Computational Intelligence: An International Journal 11(2) (1995)
324. Ziarko, W.: Special issue, Fundamenta Informaticae 27(2-3) (1996)
325. Ziarko, W.: Probabilistic decision tables in the variable precision rough set model. Computational Intelligence 17, 593–603 (2001)
326. Ziarko, W.P., Yao, Y. (eds.): RSCTC 2000. LNCS (LNAI), vol. 2005. Springer, Heidelberg (2001)

Zdzisław Pawlak (1926-2006)

He was not just a great scientist – he was also a great human being.

– Lotfi A. Zadeh, April 2006

Fig. .1. Keynote talk of Professor Pawlak during the RSKD 1993 conference in Banff.

Professor Zdzisław Pawlak is the founder of the Polish school of Artificial Intelligence and one of the pioneers in Computer Engineering and Computer Science with worldwide influence. He was truly great scientist, teacher and a human being.

Zdzisław Ignacy Pawlak was born on 10 November 1926 in Łódź, where he also finished primary school in 1939. During the German occupation of Poland, like many Poles, he was a slave-laborer and was forced to work for Siemens company. After WW II, in 1946, he passed his high school exams as an extern and, in 1947, started his studies at the Electrical Engineering Faculty of the Łódź University of Technology. In 1949, he transferred to the Electrical Faculty (the Faculty of Telecommunication between 1951–1966, the Faculty of Electronics and Information Technology at present) at the Technical University of Warsaw (now the Warsaw University of Technology). He received his engineering degree in Telecommunications and Master of Science degree in Radio Engineering in 1951, presenting the diploma thesis entitled *A clock for the electronic computing machine*, prepared under supervision of Romuald Marczyński.

After graduation, he worked as a junior member of the research staff at the Mathematical Institute of Polish Academy of Sciences (PAS) (now Institute of Mathematics of Polish Academy of Sciences (PAS)) until 1957. Between 1957–1959 he worked at the Technical University of Warsaw, where he took part in designing the first Polish computer. In effect, one of the first computing machines in Poland was built under his supervision.

In 1957, he returned to the Institute of Mathematics of PAS where he worked as an assistant professor from 1959 to 1963. He received his doctoral degree in 1958 (at the time called *candidate of technical sciences*) at the Institute of Fundamental Technological Research of PAS presenting the doctoral thesis entitled *Application of Graph Theory to the Decoder Synthesis*. The dissertation was supervised by Professor Krystyn Bochenek.

Professor Pawlak received his postdoctoral degree (habilitation, Dr. Sci.) in Mathematics at the Institute of Mathematics of PAS in 1963 for the dissertation entitled *Organization of Address-less Machines*. From 1963 until 1969, he worked at the Institute of Mathematics of the Warsaw University.

In 1971, he was promoted to an Associate Professor at Institute of Mathematics of PAS. Between 1971 and 1979, Professor Pawlak was the Deputy Director for Science at the Computer Center of PAS, and later, after the institute's renaming in 1976, of the Institute of Computer Science of PAS. In 1978, he was promoted to a (full) Professor in the Institute of Computer Science of PAS in Warsaw. In 1983, he was elected a corresponding member of PAS, and later, in 1991, he became a full member of the Polish Academy of Sciences. From 1979 to 1986, he was the director of the Institute of Informatics of the Silesian University of Technology. Starting from 1985, he worked at the Institute of Theoretical and Applied Informatics of PAS in Gliwice. In 1998, he also worked at the University of Applied Computer Science and Management in Warsaw. Between 1989 and 1996, he was the director of the Institute of Computer Science of the Warsaw University of Technology.

In 1950, Professor Pawlak developed the first generation computer GAM-1 at the Group of Computing Machines (GAM) at the State Institute of Mathematics in Warsaw. However, that machine was never used for practical applications. In 1951, Zdzisław Pawlak came up with a new way to generate random numbers, which was published in the prestigious *Mathematical Tables and Other Aids to Computation* (now called *Mathematics of Computation*), the oldest journal devoted to computation. It was the first ever Polish computer science work published abroad. Later, he suggested a new method for representing numbers in the positional numerical system with a negative radix -2 (so called -2 system). Based on this technique and horizontal microprogramming, with Professor Pawlak's project and supervision, a computing machine UMC-1 was built at the Warsaw University of Technology. Later, Professor Pawlak was studying many aspects of computer science, including computational linguistics, automata theory, automated theorem proving, and information retrieval. One of the most interesting achievements of that period was a new formal model of computing machine, different from Turing's machine and Rabin-Scott's finite automata. That model gained a lot of attention worldwide and was called Pawlak machine in the literature. Another important accomplishment was creating the first mathematical model of Crick and Watson's DNA encoding. Pawlak also developed an original approach to the information retrieval. He also proposed a new mathematical approach to the conflict analysis, which has an important influence in psychology, economy and politics.

Professor Pawlak's most important discovery was his invention of the rough set theory in 1982, which gained vast popularity throughout the World. More than 5000 English-language publications about Pawlak's theory and its applications have been published so far, including several books and over 5000 Chinese-language publications including more than 20 books.

Many international conferences, mainly in the USA, Canada, China, India, Japan and Europe were organised to discuss and develop Professor Pawlak's work. The rough sets theory has an immense following in China. Research on rough sets is significantly growing in India, recently. The year 2009 was called the Rough Set Year in India. There are also a lot of teams working on rough set theory and its applications at many university centres in Poland[1]. Professor Pawlak's book about rough sets was, so far, quoted above 8500 times in Google Scholar. The number of valuable theoretical publications and applications basing on rough sets is constantly growing, especially when combined with other approaches to reasoning based on imperfect (often incomplete) information.

At those conferences he gave lectures, among other subjects, in mathematical logic, mathematical foundations of computer science, organization of computing machines, mathematical linguistics and rough set theory. He was frequently invited as a Visiting Professor to many universities in the USA, Canada and Europe, including Philosophy Department at Stanford University (1965).

Professor Pawlak received many honours and awards acknowledging his achievements as one of the main animators of scientific life in Poland. His work was recognized on the national level by the Polish government, including Polish National Science Award in 1973, Polish Knight's Cross of the Order of Polonia Restituta in 1984, Polish Mathematical Society Steinhaus Prize for achievements in applications of Mathematics (1989) and Polish Officer's Cross of the Order Polonia Restituta in 1999.

Professor Pawlak was a member and officer of many scientific organizations, active in various periods of time in over 30 governing councils (including being the president of a number of those). In his native Poland, he was the president of the National Central Committee for Scientific Titles and Degrees (CKK) between 1975

[1] See `http://rsds.univ.rzeszow.pl/`

and 1988 (mathematical and technical sections) [2], member of the Computer Science Committee of PAS, the Committee of Cooperation of Academies of Sciences of the Socialist Countries on Computational Technology (1971 - 1979), the State Committee for Scientific Research (1994 - 2000), the Central Committee for Scientific Titles and Degrees (2000 - 2006), the Polish Mathematical Society and the Polish Semiotic Society (vice-president, 1990 - 1996). He served on several editorial boards of scientific journals, both foreign and national. He served as the deputy editor-in-chief of the Bulletin of PAS. On his initiative the journal Fundamenta Informaticae was created. For many years, he served as the deputy editor-in-chief of Fundamenta Informaticae. He published over two hundred articles and a number of books. Professor Pawlak supervised thirty doctoral dissertations. We quote all this facts to show the amount of his energy and enthusiasm devoted to promotion of scientific research, education of young researchers and their supervision.

Professor Pawlak inspired many computer scientists and mathematicians both in Poland and throughout the world. At present his students and collaborators created research teams in many countries, including, besides of his native Poland United States, Canada, Japan, Norway, Sweden and other places. It would be hardly possible to find a computer science institution, in his native Poland without encountering faculty influenced by Pawlak. Some research centers for instance in Warsaw, Poznań, Gdańsk, Katowice, Wrocław, and Rzeszów were formed following his initiative. His scientific achievements continue to inspire his many students still working in these institutions and also the next generations of students. Professor Pawlak had an unusual gift to inspire his interlocutors. As a consequence many individuals were profoundly influenced by his interests and enthusiasm towards scientific research right from the first contact with him. Let us recall here some sentences from a commemorative letter by Andrzej Skowron:

> *In middle of 1960s of the previous century, Professor Pawlak and Professor Helena Rasiowa conducted a research seminar on automated theorem proving, at the Faculty of Mathematics and Mechanics of the University of Warsaw. I remember this as if it was today. It was a big auditorium filled with participants of the seminar. In that period, Professor Pawlak was involved in an intensive research on mathematical models of computers and computations realised by them. He lectured and conducted seminars for students of mathematics and computer science. The cooperation with Professor Helena Rasiowa and her team began in the early 1960s and lasted for dozens of years. The results of this collaboration are still important. One may surely say that a new research school came into existence [2]. This cooperation had a lot of influence on shaping many people's scientific research and on evolution of the main notions of logic researched by Professor Rasiowa's group [2]: from the classical logic to non-classical logic and its inference processes, which is the main characteristics of current Artificial Intelligence investigations.*

[2] This committee was responsible for the scientific evaluation of candidates applying for D. Sci. (habilitation) degree and Professor title and recommending the final decision.

Professor Pawlak was a lively and witty person. He felt great among other people, especially friends, and adored nature. He had many interests and had many talents, for instance he sang very well and he knew probably all operetta arias by heart. He was also a gifted painter of scenes from nature.

After he passed away on April 7, 2006, many articles about his life appeared in the literature (e.g., [1-9]). In the present note, we would like to share with the reader a few reminiscences about Professor Zdzisław Pawlak by some of his close friends.

> *The road which led Professor Pawlak to his crucial discoveries was long but ended successfully. Over fifty years, Professor Pawlak researched many areas of computer science. With no hesitation, one can say that his personal way is one of the most important ones from the fifty-year-old history of research in Polish and worldwide computer science*[3].

> *The problem of finding the relevant information is one of the most urgent tasks of Computer Science. Professor Pawlak was one of the giants who created the theory that underlies the digital revolution. Rough Sets is one of the leading paradigms for thinking about the information, as it is provided to us at the global village through the World-Wide Web*[4].

> *Professor Pawlak was an unusually gifted scientist whose work is characterized by two main features: genuine originality and elegant simplicity. He had a gift of getting to the real essence, the root, of a research problem. Then he was able to formulate a model that was capturing this root in an elegant and transparent way. His real research interests were always on the boundary of applications and theory: he formulated theoretical models of phenomena that were highly relevant for applications. This reflected well his engineering background. His research was often pioneering - a good example is his model of the structure and functionality of DNA, which he formulated already in 1965. He was an unusually modest scientist - this modesty was totally disproportional to his scientific achievements. He was always amazed that his ideas had such a broad and profound influence. Even when he was describing to me the state-of-the-art in rough sets, he preferred to talk about the work of others. Usually during such discussions it took me some time to figure out that many of the nice ideas really originated with him. He was a great scientist, certainly the most influential Polish computer scientist. The combination of originality, creativity, and passion for research on the one hand and such a disarming modesty on the other made him really a role model for scientists. He was a great scientist, but he was also a wonderful person. He had a great sense of humor and a very contiguous laugh - our sessions ended often in attacks of hickups invoked by telling jokes and writing funny rhymes and poetry. The essence of what I want to say is that he was a great scientist and a wonderful human*

[3] From a commemorative talk by Professor Roman Słowiński [9].

[4] From a commemorative letter by Professor Victor Marek.

being of an exceptional integrity. I was really privileged to have him as a close friend. I surely miss him, I miss our phone conversations, and I often think about him. I know that many of his friends will remember him for a long time. As a scientist he will be remembered for a very long time as there is no doubt that many of his scientific ideas have a great future[5].

Zdzisław Pawlak gave an abundance of his time and energy to help others. His personality and insight had, undoubtedly, influence on many scientists around the World. He had a unique gift to inspire his students, co-workers and many scientists beyond his surroundings to do research. Professor's associates recognized his extraordinary character. Many called him "Our Papa Pawlak". He was with us only for a while. However, when we consider his talents and great achievements, we know how much he influenced us and our successes with his research work in many areas like approximate reasoning, intelligent information systems, computation models and foundations of computer science, and artificial intelligence - especially including rough set theory, molecular computing, pattern recognition, philosophy, art, and poetry. He also influenced us with his uncommonly rich personality. Many sources proved his ability to make sacrifices for others[6].

He was a very demanding person and, apart from being witty and using a specific humorous retorts, people showed him great respect for his knowledge and intelligence. He often repeated that scientific research is very hard and sometimes he would prefer to be a lumberjack, who may rest after the work, being surrounded with a beautiful nature. He seemed indestructible. However, his struggle with severe illness was very hard. When I was afraid that he was getting tired during our meetings and ask him if anything hurts him, he replied: "Let's not talk about it, others suffer more despite the fact they are better than me (like the Holy Father John Paul II)". Only at the end, there were moments, when he said: "You have better connections up there[7]*, tell them to take me now". Nature often lured him to primeval forests and lakes. He documented it in his beautiful photographs and painted pictures. I think that he rests somewhere among his favourite forests and lakes now but I sometimes miss his words "I haven't seen you for some time, you are getting insubordinate recently Madame Professor"*[8].

Acknowledgements

I would like to express deepest gratitude to many Colleagues who supported in different ways the preparation of this note and an extended version of it which will appear in another the book by Springer dedicated to the memory of Professor Zdzisław

[5] From a commemorative letter by Professor Grzegorz Rozenberg.
[6] From a commemorative letter by Professors James Peters and Andrzej Skowron.
[7] i.e., in heaven.
[8] From commemorative letter by Professor Alicja Wakulicz-Deja [9].

Pawlak. In particular, I would like to thank to Mohua Banerjee, Piotr Dembiński, Jerzy Fiett, Jerzy Grzymała-Busse, Mihir Kumar Chakraborty, Jerzy Dembczyński, Anna Gomolińska, Wojciech Gomoliński, Andrzej Jankowski, Solomon Marcus, Victor Marek, Mikhail Moshkov, Toshinori Munakata, Hung Son Nguyen, Hiroakira Ono, Sankar K. Pal, James Peters, Zbigniew W. Ras, Grzegorz Rozenberg, Roman Słowiński, Janusz Sosnowski, Urszula Stańczyk, Zbigniew Suraj, Marcin Szczuka, Dominik Ślęzak, Alicja Wakulicz-Deja, Guoyin Wang, Lotfi A. Zadeh.

Andrzej Skowron

Institute of Mathematics, The University of Warsaw

References

1. Ehrenfeucht, A., Peters, J. F., Rozenberg, G., Skowron, A.: Zdzisław Pawlak life and work (1926-2006). Bulletin of the European Association for Theoretical Computer Science (EATCS) **90**, 202–204 (2006)
2. Jankowski, A., Skowron, A.: Logic for artificial intelligence: The Rasiowa – Pawlak school perspective. In: Ehrenfeucht, A., Marek, V., Srebrny, M. (eds.), Andrzej Mostowski and Foundational Studies, pp. 106–143, IOS Press, Amsterdam (2008)
3. Peters, J. F., Skowron. A.: Some contributions by Zdzisław Pawlak. In: Wang, G., Peters, J.F., Skowron, A., Yao, Y. (eds.): Proceedings of the 1st International Conference on Rough Sets and Knowledge Technology (RSKT'2006), Chongqing, China, July 24-26, 2006, *Lecture Notes in Artificial Intelligence*, vol. 4062, pp. 1–11, Springer-Verlag, Heidelberg (2006)
4. Peters, J. F., Skowron. A.: Zdzisław Pawlak life and work (1926-2006). Transactions on Rough Sets V: Journal Subline, *Lecture Notes in Computer Science* **4100**, 1–24 (2006)
5. Peters, J. F., Skowron. A.: Zdzisław Pawlak Commemorating His life and work. In: Greco, S., Hata, Y., Hirano, S., Inuiguchi, M., Miyamoto, S., Nguyen, H.S., Słowiński, R. (eds.): Proceedings of the 5th International Conference on Rough Sets and Current Trends in Computing (RSCTC'2006), Kobe, Japan, November 6-8, 2006, *Lecture Notes in Artificial Intelligence*, vol. 4259, pp. 49-52, Springer-Verlag, Heidelberg (2006)
6. Peters, J. F., Skowron. A.: Zdzisław Pawlak life and work (1926-2006). Information Sciences **171** (1), 1–2 (2007)
7. Skowron, A. (ed.): New frontiers in scientific discovery – Commemorating the life and work of Zdzisław Pawlak. Fundamenta Informaticae **75** (1-4), 1-562 (2007)
8. Słowiński, R.: Prof. Zdzisław Pawlak (1926-2006). Obituary. Fuzzy Sets and Systems **157**, 2419-2422 (2006)
9. Wakulicz-Deja, A. (ed.): Decision Support Systems volume I. University of Silesia, Katowice (2007), pp. 1–206 (in Polish)

Part III

Logical Analysis of Data

3

Logical Analysis of Data: Theory, Methodology and Applications

3.1 Introduction

Logical analysis of data (LAD) is a data analysis methodology which combines ideas and concepts from optimization, combinatorics and Boolean functions. The idea of LAD was first described by Peter L. Hammer in a lecture given in 1986 at the International Conference on Multi-attribute Decision Making via OR-based Expert Systems [41] and was later expanded and developed in [32]. That first publication was followed by a stream of research studies many of which can be found in the list of references. In early publications the focus of research was on theoretical developments and on computational implementation. In recent years attention was concentrated on practical applications varying from medicine to credit risk ratings. The purpose of the present chapter is to provide an overview of the theoretical foundations of this methodology, to discuss various aspects of its implementation and to survey some of its numerous applications. We start with an introductory example proposed in [32].

A physician would like to find out the combination of food items which cause a headache to one of his patients, and requests his patient to keep a record of his diet. One week later, the patient returns to the doctor and brings in the record displayed in Table 3.1.

After a brief examination, the doctor concludes that on the days when the patient had no headache, he never consumed food #2 without food #1, but he did so on some of the occasions when he had a headache. Similarly, our clever doctor concludes that the patient has never consumed food #4 without food #6 on the days when he had no headache; but he did so once, and he had a headache. He finally concludes that the two "patterns" noticed above explain every headache, and he puts forward the "theory" that this patient's headaches can always be explained by using these two patterns.

Obviously, the doctor had to answer three questions in the process.

(a) How to come up with a short list of food items sufficient to explain the presence or absence of headaches? In our example, foods #1, 2, 4, 6 were already enough for this purpose.

I. Chikalov et al.: Three Approaches to Data Analysis, ISRL 41, pp. 147–192.
springerlink.com

(b) How to detect patterns (i.e. combinations of food items) causing headaches? In our case, the doctor found two such patterns.
(c) How to build theories (i.e. collection of patterns) explaining every observed headache?

Table 3.1. Introductory example – diet record

	Food item								
Day	1	2	3	4	5	6	7	8	Headache
1		x		x		x	x		Yes
2	x		x		x		x		No
3				x	x	x			No
4	x	x		x		x		x	No
5	x	x		x	x			x	Yes
6			x		x		x		No
7		x	x		x			x	Yes

These three questions capture the essence of LAD methodology. In what follows, we discuss these and various related questions at a more abstract level. For those who are familiar with Boolean terminology it is clear that Table 3.1 is nothing but a partially defined Boolean function on a set of 8 variables. For those who are not familiar with this terminology, we define it in Section 3.2. This section describes the main ideas of LAD in purely theoretical terms and also reviews many of the theoretical problems related to this methodology which have been studied in the literature. Then in Section 3.3 we turn to implementation of various aspects of this methodology. Finally, in Section 3.4 we illustrate the power of LAD with a number of specific applications.

3.2 Theory

The central notions of LAD are that of a partially defined Boolean function and of a pattern. In this section, we define these and related notions and consider various problems associated with them.

3.2.1 Terminology and Notation

A binary word of length n is an ordered sequence of 0's and 1's of length n. If $\alpha_1\alpha_2\ldots\alpha_n$ is a binary word, then α_i is called the i-th letter (or i-th component) of the word. Since each component of a binary word can take one of the two possible values, there are 2^n binary words of length n. On the set of binary words of the same length we define a partial order $\leq$ as follows: $\alpha = (\alpha_1\ldots\alpha_n) \leq \beta = (\beta_1\ldots\beta_n)$ if and only if $\alpha_i \leq \beta_i$ for each $i = 1,\ldots,n$.

Let us denote $B = \{0,1\}$. The set B^n consists of all binary words of length n and is commonly referred to as the *Boolean hypercube* of dimension n. For instance, the set B^3 consists of the eight binary sequences of length 3 listed in the first column of Table 3.2. Sometimes, the elements of B^n are called 0-1 vectors or simply points of the hypercube.

Table 3.2. A Boolean function of three variables

x_1 x_2 x_3	$f(x_1,x_2,x_3)$
0 0 0	1
0 0 1	0
0 1 0	1
0 1 1	1
1 0 0	0
1 0 1	0
1 1 0	1
1 1 1	0

A subset $S \subseteq B^n$ is a *subcube* of B^n if $|S| = 2^k$ for some $k \leq n$ and there are $n-k$ components in which all sequences of S coincide. The number k is called the dimension of S. For instance, the first four sequences of Table 3.2 create a subcube of dimension 2; they all coincide in the first component. A single point of B^n is a subcube of dimension 0.

A *Boolean function* of n variables $x_1,\dots,x_n$ is a mapping $f : B^n \to B$. Table 3.2 represents an example of a Boolean function of three variables. To each of the 2^n binary vectors, a Boolean function assigns one of the two possible values. Therefore, there are 2^{2^n} Boolean functions of n variables.

Given a Boolean function f, a binary vector $\alpha = (\alpha_1\alpha_2\dots\alpha_n)$ is called a *true point* of the function if $f(\alpha) = 1$ and a *false point* if $f(\alpha) = 0$. The sets of true and false points of a function f will be denoted by $T = T(f)$ and $F = F(f)$, respectively. For instance, for the function f represented in Table 3.2, $T(f) = \{000,010,011,110\}$ and $F(f) = \{001,100,101,111\}$.

If x is a Boolean variable (i.e. variable taking values 0 and 1), then $\overline{x} = 1-x$ is the *complement* (or *negation*) of x. Both the variables and their complements are called *literals*. Sometimes, variables are called *positive* literals and their complements are called *negative* literals. An *elementary conjunction* (or simply, a *term*) is a product of literals. The *degree* of a term is the number of literals in it. For instance, the degree of the term $x_2\overline{x}_3$ is 2, and the degree of $\overline{x}_1x_2x_3$ is 3. Obviously, every term can be viewed as a Boolean function, i.e. it maps 0-1 vectors to $\{0,1\}$. Consider, for instance, the term $t = \overline{x}_1x_2x_3$, then $t(0,0,1) = \overline{0}\cdot 0\cdot 1 = 0$. Notice that the value of $t(\alpha)$ is defined for each 0-1 vector $\alpha \in \{0,1\}^n$ even if the degree of t is less than n, we simply ignore the values of the variables which are not present in the term either as positive or as negative literals. For instance, if $t = x_2\overline{x}_3$ and $\alpha = (0,0,1)$, then

$t(\alpha) = 0 \cdot \overline{1} = 0$, since the first component of α corresponds to the variable x_1. A term t is said to *cover* a point $\alpha \in \{0,1\}^n$ if $t(\alpha) = 1$.

There is a close connection between terms and subcubes of B^n, which is reflected in the following proposition.

Proposition 3.1. *A subset S of B^n is a subcube of dimension k if and only if there is a term t of degree $n-k$ such that the set of points of B^n covered by t is S.*

This proposition allows us to use the two notions, subcubes and terms, interchangeably. To illustrate it, let us again consider the subcube of B^3 formed by the first four sequences of Table 3.2. This subcube has dimension 2 and the unique term covering it is $\overline{x}_1$.

The *characteristic term* of a point $\alpha \in \{0,1\}^n$ is the unique term of degree n covering α. Characteristic terms are also known as *minterms*. For instance, the characteristic term of $(0,1,1)$ is $\overline{x}_1 x_2 x_3$.

Let f be a Boolean function of n variables. A term t is an *implicant* of f if $t(\alpha) \leq f(\alpha)$ for each $\alpha \in \{0,1\}^n$. In other words, t is an implicant of f if the set of the points of B^n covered by t is a subset of the true points of f. An implicant t is called *prime* if it is minimal, i.e. deletion of any literal from t results in a term which is not an implicant.

Every partition of the set B^n of all 0-1 n-vectors into two disjoint sets T and F defines a Boolean function on B^n. Now assume that the sets T and F are disjoint but cover B^n not entirely, i.e. some points of B^n belong neither to T nor to F. Then we have a function which is defined only partially, a *partially defined Boolean function* (pdBf). This function is given by:

$$f(\alpha) = \begin{cases} 1 \text{ if } \alpha \in T \\ 0 \text{ if } \alpha \in F \end{cases}$$

A function f defined on a set of true points T and a set of false points F will be denoted $f = (T,F)$.

Table 3.3. A partially defined Boolean function

	x_1	x_2	x_3	x_4	x_5	x_6	x_7	x_8	$f(x_1,\ldots,x_8)$
1	0	1	0	1	0	1	1	0	1
2	1	0	1	0	1	0	1	0	0
3	0	0	0	1	1	1	0	0	0
4	1	1	0	1	0	1	0	1	0
5	1	1	0	1	1	0	0	1	1
6	0	0	1	0	1	0	1	0	0
7	0	1	1	0	1	0	0	1	1

Table 3.3 gives an example of a partially defined Boolean function of eight variables. This function is defined only on 7 points of the hypercube B^8 numbered 1 through 7. The points 1,5,7 are the true points of the function and 2,3,4,6 are its false points. An attentive reader can easily recognize in this function the diet record of Table 3.1. This table can be extended by adding to it new records. Similarly, every partially defined Boolean function can be extended by defining its values on new points of the hypercube. Every Boolean function (i.e. defined over all 2^n 0-1 vectors) agreeing with a pdBf $f = (T,F)$ on $T \cup F$ and taking arbitrary 0-1 values elsewhere is called an *extension* of f. The number of extensions can be very large. In particular, if a function f is defined on k points of the hypercube B^n, then there exist 2^{2^n-k} ways to extend it. Among these many extensions, LAD aims at distinguishing a "right" one. There is no definition for a "right extension". However, we assume that any real-life dataset is not just a collection of random facts and any rational phenomenon (headache, etc.) has a rational explanation. The idea of LAD is to learn this explanation from the partial information at hand. LAD does this job in two major steps. The first step is

(1) Detecting patterns

Pattern is the key notion of the LAD methodology. In Section 3.2.2, we define this notion, consider different types of patterns, and discuss their relative efficiency. Then in Section 3.3.3 we return to this notion and describe various algorithms for detecting patterns.

The first step can produce a large amount of patterns not all of which may be relevant to explaining the phenomenon under study, in which case LAD proceeds to the second step:

(2) Theory formation

A *theory*, also called a *model*, is an extension of a partially defined Boolean function. Since the number of extensions can be very large, it is important to devise additional principles to help in identifying particularly appealing ones. Several specific types of extensions studied in the literature are discussed in Section 3.2.3 and implementation details of this step are discussed in Section 3.3.4.

3.2.2 Patterns

Let $f = (T,F)$ be a partially defined Boolean function. A term t is a *positive pattern* of f if it covers at least one true point and no false point of f. Alternatively, a *positive pattern* is a subcube of B^n that intersects T and disjoint from F. *Negative patterns* are defined in a similar way. Since the properties of positive and negative patterns are completely symmetric, without loss of generality we will focus on positive patterns and will frequently refer to positive patterns simply as patterns.

Patterns play a key role in LAD, since they admit a clear interpretation by human experts. Consider, for instance, the partially defined Boolean function of Table 3.3, which models the diet record of the introductory example. The term $\overline{x}_1 x_2$ equals 1 if and only if $x_1 = 0$ and $x_2 = 1$, therefore it covers points 1 and 7 and does not cover any other point of the table. Since 1 and 7 are the true points of the function, we conclude that $\bar{x}_1 x_2$ is its positive pattern. This pattern suggests a special role of food 2 ($x_2 = 1$) consumed without food 1 ($x_1 = 0$). Similarly, $x_4 \bar{x}_6$ is a positive pattern. The only point which is covered by this term is point 5, and this is a true point. Below we shall see that the pdBf of Table 3.3 has many more patterns.

Typically, a partially defined Boolean function has exceedingly many patterns and generation of all of them is computationally expensive. In addition, it has been observed in empirical studies and practical applications that some patterns are more "suitable" than others for use in data analysis. Unfortunately, the concept of suitability does not have a unique definition. Among the many reasonable criteria of suitability, paper [44] distinguishes three - simplicity, selectivity and evidence - and we discuss them in Section 3.2.2.1. Then in Section 3.2.2.2 we introduce one more pattern type - maximum patterns - and describe a linear integer program for finding them. In Section 3.2.2.3 we discuss a mixed 0-1 integer and linear programming approach to identifying LAD patterns that are optimal with respect to various criteria. In Section 3.2.2.4 we extend the notion of a pattern to that of an interval, which is an important tool in case of non-binary datasets.

3.2.2.1 Pareto-optimal Patterns in Logical Analysis of Data

We start by introducing some terminology of the theory of partially ordered sets. Given a set A, we denote by A^2 the set of all ordered pairs (x,y) such that $x \in A$ and $y \in A$. A binary relation on A is a subset of A^2. We will say that two elements $x,y \in A$ are comparable with respect to a binary relation ρ if either $(x,y) \in \rho$ or $(y,x) \in \rho$, and incomparable otherwise.

A binary relation ρ on A is called a *partial preorder* (also known as *quasi-order*) if it is

- *reflexive*, i.e. $(x,x) \in \rho$ for all $x \in A$,
- *transitive*, i.e. $(x,y) \in \rho$ and $(y,z) \in \rho$ implies $(x,z) \in \rho$.

A partial preorder is called a *partial order* if it is also

- *antisymmetric*, i.e. $(x,y) \in \rho$ and $(y,x) \in \rho$ imply $x = y$.

We will consider partial (pre)orders on the set of patterns. Let f be a partially defined Boolean function of n variables and P a pattern of f. We will denote by

- $Lit(P)$ the set of literals in P,
- $S(P)$ the subcube of P, i.e. the set of points of B^n covered by P,
- $Cov(P)$ the set of true points of f covered by P, called *coverage* of P.

Now with these three parameters we associate three partial preorders, called *simplicity preference, selectivity preference*, and *evidential preference*.

Definition 3.2.

- *Simplicity preference* σ *is a binary relation on the set of patterns such that* $(P_1,P_2) \in \sigma$ *(also denoted* $P_1 \preccurlyeq_\sigma P_2$*) if and only if* $Lit(P_1) \supseteq Lit(P_2)$.

- *Selectivity preference* Σ *is a binary relation on the set of patterns such that* $(P_1, P_2) \in \Sigma$ *(also denoted* $P_1 \preccurlyeq_\Sigma P_2$*) if and only if* $S(P_1) \supseteq S(P_2)$.
- *Evidential preference* ε *is a binary relation on the set of patterns such that* $(P_1, P_2) \in \varepsilon$ *(also denoted* $P_1 \preccurlyeq_\varepsilon P_2$*) if and only if* $Cov(P_1) \subseteq Cov(P_2)$.

For any preference $\preccurlyeq$, the simultaneous satisfaction of the relations $P_1 \preccurlyeq P_2$ and $P_2 \preccurlyeq P_1$ will be denoted $P_1 \approx P_2$. Also, the simultaneous satisfaction of the relations $P_1 \preccurlyeq P_2$ and $P_2 \not\preccurlyeq P_1$ will be denoted $P_1 \prec P_2$. For partial orders, the relation $P_1 \approx P_2$ coincides with $P_1 = P_2$. For instance, it is not difficult to see that if for two patterns P_1 and P_2 we have $Lit(P_1) \subseteq Lit(P_2)$ and $Lit(P_1) \supseteq Lit(P_2)$, then $P_1 = P_2$. Similarly, $S(P_1) \subseteq S(P_2)$ and $S(P_1) \supseteq S(P_2)$ imply $P_1 = P_2$. Therefore, σ and Σ are partial orders. This is not true for the evidential preference, since there may exist two different patterns covering the same set of true points of f. For instance, we know that the term $x_4\overline{x}_6$ is a positive pattern of the partially defined Boolean function of Table 3.3 and the only true point covered by this term is point 5. Similarly, the term $\overline{x}_3x_5\overline{x}_6$ is a positive pattern of this function and it also covers only point 5. Therefore, $x_4\overline{x}_6 \preccurlyeq_\varepsilon \overline{x}_3x_5\overline{x}_6$ and $\overline{x}_3x_5\overline{x}_6 \preccurlyeq_\varepsilon x_4\overline{x}_6$, i.e. $x_4\overline{x}_6 \approx_\varepsilon \overline{x}_3x_5\overline{x}_6$.

In order to better illustrate the above preferences, let us consider more examples. The term $x_5\overline{x}_6\overline{x}_7x_8$ covers a point of a hypercube if and only if $x_5 = 1, x_6 = 0, x_7 = 0, x_8 = 1$. Therefore, it covers two positive points of Table 3.3 (namely, points 5 and 7) and no negative points, and hence it is a positive pattern of the pdBf defined by this table. Similarly, the term $x_5\overline{x}_6x_8$ is a positive pattern. Since $Lit(x_5\overline{x}_6\overline{x}_7x_8) \supset Lit(x_5\overline{x}_6x_8)$, we have $x_5\overline{x}_6\overline{x}_7x_8 \prec_\sigma x_5\overline{x}_6x_8$, i.e. $x_5\overline{x}_6x_8$ is better than $x_5\overline{x}_6\overline{x}_7x_8$ in terms of the simplicity preference. On the other hand, $x_5\overline{x}_6x_8$ and $x_5\overline{x}_6\overline{x}_7x_8$ cover the same set of true points, and therefore, $x_5\overline{x}_6\overline{x}_7x_8 \approx_\varepsilon x_5\overline{x}_6x_8$. Also, we know that $x_4\overline{x}_6$ is a positive pattern covering only point 5, therefore, $x_4\overline{x}_6 \prec_\varepsilon x_5\overline{x}_6x_8$ and $x_4\overline{x}_6 \prec_\varepsilon x_5\overline{x}_6\overline{x}_7x_8$. However, the term $x_4\overline{x}_6$ is not comparable to $x_5\overline{x}_6x_8$ and to $x_5\overline{x}_6\overline{x}_7x_8$ with respect to the simplicity or selectivity preferences.

The three preferences defined above are not independent of each other. In particular, the following three relationships can be easily seen to hold:

(1) $P_1 \preccurlyeq_\sigma P_2$ if and only if $P_2 \preccurlyeq_\Sigma P_1$,
(2) if $P_1 \preccurlyeq_\sigma P_2$, then $P_1 \preccurlyeq_\varepsilon P_2$,
(3) if $P_1 \preccurlyeq_\Sigma P_2$, then $P_2 \preccurlyeq_\varepsilon P_1$.

In order to identify the most suitable types of patterns, paper [44] considers combinations of the preferences introduced above.

Definition 3.3. *Given two preferences* π *and* ρ *on the set of patterns,*

- *a pattern* P_1 *is preferred to a pattern* P_2 *with respect to the intersection* $\pi \cap \rho$ *if and only if* $P_2 \preccurlyeq_\pi P_1$ *and* $P_2 \preccurlyeq_\rho P_1$.
- *a pattern* P_1 *is preferred to a pattern* P_2 *with respect to the lexicographic refinement* $\pi|\rho$ *if and only if either* $P_2 \prec_\pi P_1$ *or* $P_1 \approx_\pi P_2$ *and* $P_2 \preccurlyeq_\rho P_1$.

The preferences $\pi \cap \rho$ and $\rho \cap \pi$ are identical for any π and ρ, while $\pi|\rho$ and $\rho|\pi$ are usually different. Also, if π is a partial order, then the lexicographic refinement $\pi|\rho$ is identical to π for any ρ. Therefore, the lexicographic refinements $\sigma|\varepsilon$ and

$\Sigma|\varepsilon$ coincide with σ and Σ, respectively. Also, because of the relationship (1), any combination of the simplicity and selectivity preferences using either intersection or lexicographic refinement makes no sense, and according to (2), the intersection $\sigma \cap \varepsilon$ is identical to σ.

In conclusion, the only new preferences that can be obtained by applying intersection and lexicographic refinement are $\Sigma \cap \varepsilon$, $\varepsilon|\sigma$ and $\varepsilon|\Sigma$.

Definition 3.4. *Given a preference $\preccurlyeq$ on the set of patterns, a pattern P will be called Pareto-optimal with respect to $\preccurlyeq$ if there is no pattern P' different from P such that $P \prec P'$.*

Definition 3.5. *A pattern which is Pareto-optimal with respect to*

- *the simplicity preference is called* PRIME. *In other words, a patter is prime if and only if the removal of any literal from $Lip(P)$ results in a term which is not a pattern.*
- *the selectivity preference is the characteristic term (minterm) of a true point.*
- *the evidential preference is called* STRONG. *In other words, a pattern P is strong if there is no pattern P' such that $Cov(P) \subset Cov(P')$.*
- *$\Sigma \cap \varepsilon$ is called* SPANNED.

The patterns that are optimal with respect to $\varepsilon|\sigma$ and $\varepsilon|\Sigma$ do not require special names, since they can be described using already introduced terms, which follows from the theorem below proved in [44].

Theorem 3.6. *A pattern is Pareto-optimal with respect to*

- *$\varepsilon|\sigma$ if and only if it is both strong and prime,*
- *$\varepsilon|\Sigma$ if and only if it is both strong and spanned.*

Table 3.4 below summarizes the properties of patterns which are Pareto-optimal with respect to the preferences and combinations of preferences discussed above.

Table 3.4. Types of Pareto-optimality

Preference	Pareto-optimal pattern
σ	Prime
Σ	Minterm
ε	Strong
$\Sigma \cap \varepsilon$	Spanned
$\varepsilon\|\sigma$	Strong prime
$\varepsilon\|\Sigma$	Strong spanned

To illustrate the notion of Pareto-optimality, let us return to the example of the partially defined Boolean function of Table 3.3. We know that the term $x_5\bar{x}_6x_8$ is a positive pattern of this function and the set of true points covered by it consists of points 5 and 7. In order to see if it is prime, let is try to obtain a shorter term by deleting one of its literals. By deleting x_5 we obtain the term $\bar{x}_6 x_8$, which covers the negative point 4 and hence is not a positive pattern anymore. Similarly, by deleting x_8 we obtain a term which is not a pattern of the function. However, the term $x_5 x_8$ obtained by deleting $\bar{x}_6$ is a positive pattern covering points 5 and 7. Therefore, $x_5\bar{x}_6x_8 \prec_\sigma x_5x_8$ and hence $x_5\bar{x}_6x_8$ is not optimal with respect to the simplicity preference, i.e. it is not prime. The term x_5x_8, however, is prime since the deletion of any literal from it results in a term which is not a pattern. This term is also strong, simply because there are no patterns covering more than two true points, which can be easily verified. It is, however, not spanned, since it is not optimal with respect to $\Sigma \cap \varepsilon$. Indeed, the $x_5\bar{x}_6x_8$ covers the same set of true points as x_5x_8, but $S(x_5\bar{x}_6x_8) \subseteq S(x_5x_8)$. With a bit of work the reader can find out that $x_2x_5\bar{x}_6\bar{x}_7x_8$ is a spanned pattern, i.e. we cannot increase its selectivity (by adding more literals to the pattern) without decreasing the coverage.

Finally, we observe that the term $\bar{x}_1x_2$, which is a positive pattern, is prime, strong and spanned.

3.2.2.2 Maximum Patterns

For a binary vector α, an α-pattern is a pattern covering α. A *maximum* α-pattern is an α-pattern P with maximum coverage, i.e. with maximum number of positive points covered by P (if α is positive) or with maximum number of negative points covered by P (if α is negative). Remember that by definition P cannot cover both a positive and a negative point.

The notion of maximum patterns was introduced in [19], where the authors also proposed an integer program for the problem of constructing a pattern of maximum coverage which includes a given point. In view of the perfect symmetry of positive and negative patterns, we will describe the proposed program only for the positive case.

Let $f = (T,F)$ be a partially defined Boolean functions of n variables $x_1,\dots,x_n$ and $\alpha = (\alpha_1,\dots,\alpha_n)$ a positive point. We repeat that a pattern is a conjunction of literals. If a pattern covers α and contains a variable x_i, then this variable appears in the pattern as a positive literal if and only if $\alpha_i = 1$. Therefore, constructing a pattern P that covers α consists in deciding, for each $i = 1,\dots,n$, whether or not we include the i-th variable in P. To this end, for each $i = 1,\dots,n$, we introduced a binary decision variable y_i which equals 1 if and only if x_i is included in P.

If $\beta = (\beta_1,\dots,\beta_n)$ is another positive point covered by P and $\alpha_i \neq \beta_i$ for some i, then variable x_i is necessarily *not* in P, in which case $y_i = 0$. Therefore, the number of positive points covered by P will be given by

$$\sum_{\beta \in T} \prod_{\substack{i=1 \\ \alpha_i \neq \beta_i}}^{n} \bar{y}_i,$$

which defines the objective function of our integer program.

In order to define constraints, we observe that a pattern covering α should not cover any negative point $\gamma = (\gamma_1, \ldots, \gamma_n)$. We know that if $y_i = 0$ for all i such that $\alpha_i \neq \gamma_i$, then the corresponding pattern covers both α and γ. Therefore, to assure that γ is not covered by the pattern we require that

$$\sum_{\substack{i=1 \\ \alpha_i \neq \gamma_i}}^{n} y_i \geq 1. \tag{3.1}$$

Combining the objective function with the constraints we obtain the following non-linear integer program:

$$\begin{aligned} \text{maximize} \quad & \sum_{\beta \in T} \prod_{\substack{i=1 \\ \alpha_i \neq \beta_i}}^{n} \bar{y}_i \\ \text{subject to} \quad & \sum_{\substack{i=1 \\ \alpha_i \neq \gamma_i}}^{n} y_i \geq 1 \text{ for every } \gamma \in F \\ & y_i \in \{0,1\} \text{ for every } i = 1, \ldots, n. \end{aligned} \tag{3.2}$$

Computationally, this is a difficult problem, i.e. it is NP-hard, which means no polynomial-time algorithms are available to solve the problem. On the other hand, there are numerous software packages for solving *linear* integer programs. Therefore, it is useful to rewrite (3.2) as a linear integer program. To this end, for each positive point β different from α we introduce a new binary variable z_β to replace each term of the objective function of problem (3.2):

$$z_\beta = \prod_{\substack{i=1 \\ \alpha_i \neq \beta_i}}^{n} \bar{y}_i.$$

Let us denote by $w(\beta)$ the number of components where β differs from α. Then the definition of z_β is equivalent to the following two inequalities:

$$w(\beta) z_\beta \leq \sum_{\substack{i=1 \\ \alpha_i \neq \beta_i}}^{n} \bar{y}_i,$$

$$\sum_{\substack{i=1 \\ \alpha_i \neq \beta_i}}^{n} \bar{y}_i \leq z_\beta + w(\beta) - 1.$$

Also, we observe that

$$\sum_{\substack{i=1 \\ \alpha_i \neq \beta_i}}^{n} \bar{y}_i + \sum_{\substack{i=1 \\ \alpha_i \neq \beta_i}}^{n} y_i = \sum_{\substack{i=1 \\ \alpha_i \neq \beta_i}}^{n} (\bar{y}_i + y_i) = \sum_{\substack{i=1 \\ \alpha_i \neq \beta_i}}^{n} 1 = w(\beta).$$

Therefore, the non-linear integer program (3.2) can be rewritten as the following linear integer program:

$$\begin{aligned}
\text{maximize} \quad & \sum_{\beta \in T - \{\alpha\}} z_\beta \\
\text{subject to} \quad & \sum_{\substack{i=1 \\ \alpha_i \neq \gamma_i}}^{n} y_i \geq 1 \text{ for every } \gamma \in F \\
& w(\beta) z_\beta + \sum_{\substack{i=1 \\ \alpha_i \neq \beta_i}}^{n} y_i \leq w(\beta) \text{ for every } \beta \in T - \{\alpha\} \\
& z_\beta + \sum_{\substack{i=1 \\ \alpha_i \neq \beta_i}}^{n} y_i \geq 1 \text{ for every } \beta \in T - \{\alpha\} \\
& y_i \in \{0,1\} \text{ for every } i = 1, \ldots, n, \\
& z_\beta \in \{0,1\} \text{ for every } \beta \in T - \{\alpha\}.
\end{aligned} \tag{3.3}$$

The number of binary variables appearing in Problem (3.3) is $n + |T| - 1$, which in case of large datasets results in very large integer linear programs. In view of the computational difficulty of handling such large integer linear programs, it is important to develop appropriate heuristics to deal with instances for which current linear programming software packages fail to find exact solutions. Two of such heuristics are described in Section 3.3.3.3.

3.2.2.3 MILP Approach to Pattern Generation

A Mixed 0-1 Integer and Linear Programming (MILP) approach to identifying LAD patterns that are optimal with respect to various preferences has been proposed in [69]. In what follows we present an MILP formulation of the problem of identifying positive patterns. For negative patterns, the formulation is similar.

Let $f = (T, F)$ be a partially defined Boolean function and $x_1, \ldots, x_n$ the set of its variables. To simplify the description, let us denote the negation of x_i by x_{n+i}. Therefore, $x_1, \ldots, x_n, x_{n+1}, \ldots, x_{2n}$ is the set of all literals.

Since a pattern is a conjunction of literals, the problem of identifying a pattern consists in deciding, for each of the $2n$ literals, whether we include it in the pattern we are looking for or not. To this end, we introduce $2n$ decision variables $z_1, \ldots, z_n, z_{n+1}, \ldots, z_{2n}$ such that x_i is included in the pattern if and only if $z_i = 1$ for each $i = 1, \ldots, 2n$. Obviously, we do not want patterns containing both a variable and its negation, because such terms do not cover any point. Therefore, we introduce the following constraints:

$$z_i + z_{n+i} \leq 1 \text{ for each } i = 1, \dots, n. \tag{3.4}$$

We also introduce the following constraint:

$$\sum_{i=1}^{2n} z_i = d. \tag{3.5}$$

Any solution satisfying (3.4) and (3.5) is a term of degree d, which may or may not be a positive pattern. Let us repeat that a positive pattern must cover at least one positive point and no negative point. To satisfy these requirements, we introduce more constraints. Again, to simplify the description, given a 0-1 vector $\alpha = (\alpha_1 \alpha_2 \dots \alpha_n)$, we denote, for $i = 1, \dots, n$, the complement of α_i by α_{n+i}. Then the constraint $\sum_{i=1}^{2n} \alpha_i z_i \leq d - 1$ tells us that any solution satisfying this constraint, as well as constraint (3.5), does not cover the point α. Therefore, for the negative points we introduce the following constraints:

$$\sum_{i=1}^{2n} \alpha_i z_i \leq d - 1 \quad \text{for each negative point } \alpha \in F. \tag{3.6}$$

To make sure that our solution covers at least one positive point, for each $\alpha \in T$ we introduce a binary variable y_α such that $y_\alpha = 0$ if $\sum_{i=1}^{2n} \alpha_i z_i \geq d$ (i.e. if our solution covers α) and $y_\alpha = 1$ otherwise. Now, for the positive points we introduce the following constraints:

$$\sum_{i=1}^{2n} \alpha_i z_i + n y_\alpha \geq d \quad \text{for each positive point } \alpha \in T. \tag{3.7}$$

Summarizing, we obtain the following mixed 0-1 integer linear program:

$$\begin{array}{lll}
\min\limits_{\mathbf{z},\mathbf{y},d} & \sum\limits_{\alpha \in T} y_\alpha & \\
\text{subject to} & \sum\limits_{i=1}^{2n} \alpha_i z_i + n y_\alpha \geq d & \text{for each positive point } \alpha \in T \\
& \sum\limits_{i=1}^{2n} \alpha_i z_i \leq d - 1 & \text{for each negative point } \alpha \in F \\
& z_i + z_{n+i} \leq 1 & \text{for each } i = 1, \dots, n \\
& \sum\limits_{i=1}^{2n} z_i = d & \\
& 1 \leq d \leq n & \\
& \mathbf{z} \in \{0,1\}^{2n} & \\
& \mathbf{y} \in \{0,1\}^{|T|} &
\end{array} \tag{3.8}$$

The following theorem was proved in [69]:

Theorem 3.7. *Let* $(\mathbf{z}, \mathbf{y}, d)$ *be an optimal solution of* (3.8) *and* P *the term defined by* $\mathbf{z}$, *then* P *is a strong positive pattern of degree* d.

The usefulness of the program (3.8) is due to the fact that it can be easily modified to finding patterns that satisfy various other selection criteria. For instance, by substituting the objective function of (3.8) with

$$\min_{\mathbf{z},\mathbf{y},d} \sum_{\alpha\in T} y_\alpha + cd,$$

we obtain a program that finds a strong prime pattern, if $c > 0$, or strong spanned pattern, if $c < 0$.

Now let us show how the program (3.8) can be modified to finding a maximum pattern covering a positive point $\beta = (\beta_1\beta_2\dots\beta_n)$. In this case, not all literals can be used in the search for a pattern. Namely, we are limited to those literals that take value 1 at β. Let I_β the set of indices of such literals. More formally: $I_\beta = \{i = 1,\dots,2n \ : \ \beta_i = 1\}$. Clearly, the set $|I_\beta| = n$ and in each pair $(i, n+i)$ it contains exactly one of the indices. This means, in particular, that the constraint $z_i + z_{n+i} \leq 1$ is not relevant any more and can be removed from the program. By making this and some other obvious changes (changing the limits of the summations, etc) we transform (3.8) into the following program:

$$\begin{array}{lll}
\min\limits_{\mathbf{z},\mathbf{y},d} & \sum\limits_{\alpha\in T-\{\beta\}} y_\alpha & \\
\text{subject to} & \sum\limits_{i\in I_\beta} \alpha_i z_i + n y_\alpha \geq d & \text{for each positive point } \alpha \in T - \{\beta\} \\
& \sum\limits_{i\in I_\beta} \alpha_i z_i \leq d-1 & \text{for each negative point } \alpha \in F \\
& \sum\limits_{i\in I_\beta} z_i = d & \\
& 1 \leq d \leq n & \\
& \mathbf{z} \in \{0,1\}^n & \\
& \mathbf{y} \in \{0,1\}^{|T|-1} &
\end{array} \tag{3.9}$$

The following theorem was proved in [69]:

Theorem 3.8. *Let* $(\mathbf{z},\mathbf{y},d)$ *be an optimal solution of* (3.9) *and* P *the term defined by* $\mathbf{z}$, *then* P *is a maximum positive pattern of degree* d *covering* β.

3.2.2.4 Patterns vs. Intervals

Unlike the introductory example, which is artificially invented to better explain the concept of LAD, most real-life data come in a non-binary form. To overcome this difficulty, the user should either binarize the data or develop tools capable of processing non-binary data. We discuss the problem of data binarization in Section 3.3.1. In the present section, we introduce terminology extending the notion of a pattern to a more general context and mention some problems related to this extension.

Let L be an arbitrary set of numbers and $L^n = \{(x_1,\ldots,x_n) \mid x_i \in L \; \forall i = 1,\ldots,n\}$. For any two points $x = (x_1,\ldots,x_n)$ and $y = (y_1,\ldots,y_n)$ in L^n,

- we say that $x \leq y$ if $x_i \leq y_i$ for each $i = 1,\ldots,n$.
- if $x \leq y$, we call the set $I[x,y] = \{z \in L^n \mid x \leq z \leq y\}$ the *interval* (also known as the *box* [45]) defined by x and y.

In any set $S \subseteq L^n$ there is a unique minimal element $x^S_{\min}$ and a unique maximal element $x^S_{\max}$ with respect to the partial order $\leq$ defined above. For each $i = 1,\ldots,n$, the i-th coordinate of $x^S_{\min}$ is the minimum of the i-th coordinates taken over all points in S. Similarly, the i-th coordinate of $x^S_{\max}$ is the maximum of the i-th coordinates taken over all points in S. The *hull* $H[S]$ of the set S (also known as *box-closure* [45]) is the interval $H[S] := I[x^S_{\min}, x^S_{\max}]$.

In the Boolean case (when $L = \{0,1\}$), the notion of an interval coincides with the notion of a subcube. For instance, we know that the set S consisting of the first four sequences of Table 3.2 forms a subcube. The minimal element of this set is $(0,0,0)$, the maximal element is $(0,1,1)$ and the set S itself is the interval $[(0,0,0),(0,1,1)]$.

Assume that in the set L^n we are given two disjoint sets of points, called positive points and negative points. Any box (interval) containing at least one positive and no negative point is a *positive box*, and any box containing at least one negative and no positive point is a *negative box*. A box which is either positive or negative is called *homogeneous*. Clearly, the notion of positive (negative) boxes generalizes the notion of positive (negative) patterns. We also observe that hulls generalize the notion of spanned patterns.

Motivated by logical analysis of data, paper [35] studies the maximum box problem. In this problem, we are given two finite sets in $\mathbb{R}^n$ ($\mathbb{R}$ is the set of all real numbers) and would like to find a positive (negative) box containing maximum number of positive (negative) points. The authors of [35] show that the problem is generally NP-hard (though solvable in polynomial time for any fixed n), propose an integer programming formulation for the problem and an efficient branch-and-bound algorithm to solve it.

In Section 3.3.3, we will described several algorithms for generating patterns of various types. In particular, the algorithm for the generation of spanned patterns is described in the terminology of intervals and therefore can be applied to data which is not necessarily binary.

3.2.3 Theory Formation

Speaking theoretically, the ultimate goal of LAD is finding an extension of a partially defined Boolean function $f = (T,F)$. Such an extension is called a *theory* or a *model*.

The disjointness of the sets of positive points T and negative points F is a necessary and sufficient condition for the existence of an extension. Moreover, a typical data set usually has exponentially many extensions. In the absence of any additional information about the properties of the data set, the choice of an extension would be totally arbitrary, and therefore would risk to omit the most significant features of the set. However, in many practical cases significant information about the data set is available. This information can be used to restrict the set of possible extensions to those satisfying certain required properties. This leads to the following problem, where $\mathscr{C}$ denotes a (restricted) class of Boolean functions:

Problem EXTENSION($\mathscr{C}$). Given a partially defined Boolean function $f = (T,F)$, determine if there is an extension of f in the class $\mathscr{C}$.

In the computational learning theory this problem is called the *consistency problem*. Below we mention several types of Boolean functions that have been studied in the literature in relation with the Logical Analysis of Data.

Positive functions. A Boolean function f is called *positive* (or *monotonically non-decreasing*) if $\alpha \leq \beta$ implies $f(\alpha) \leq f(\beta)$. It is well-known that a Boolean function is positive if each of its prime implicants consists of only positive literals.

Positive functions have an intuitive meaning in a variety of contexts. In the example of the headache (i.e. in the introductory example), the interpretation will simply be that no food item is supposed to ever have an inhibitive effect: it either contributes (in certain combinations) to the headache or has no effect at all. More formally, a change of the value of one of the variables from 0 to 1 cannot decrease the value of the function from 1 to 0, whatever the values of the other variables are. In some cases, this fact itself may be regarded as an important information for the understanding of the mechanism causing the phenomenon.

Low-degree functions. It is known that every Boolean function can be expressed as a (not necessarily unique) disjunctive normal form (DNF), i.e. as an expression $C_1 \vee C_2 \vee \ldots \vee C_k$, where each C_i is an elementary conjunction. The degree of a DNF is the maximum of degrees of its elementary conjunctions. The degree of a Boolean function is the degree of its DNF expression(s) of lowest degree. For instance, the degree of the function given by the DNF expression $xyz \vee x\bar{v} \vee \bar{z}v$ is 2, since the same function also admits the expression $xy \vee x\bar{v} \vee \bar{z}v$ (and since it is obviously has no expression of degree 1). By analogy with the usual algebraic concept, the functions of degree 1 are called *linear* and of degree 2 *quadratic*. The interest in restricting to low-degree theories is due to the obvious fact that the lower the degree of a theory, the simpler its interpretation.

Threshold functions. A Boolean function $f(x_1,\ldots,x_n)$ is *threshold* if there exist numbers $w_1,\ldots,w_n$ and q such that $f(x_1,\ldots,x_n) = 1$ if and only if $\sum_{i=1}^{n} w_i x_i \geq q$.

Convex functions. To define the notion of a convex function, we need to introduce some more terminology. The Hamming distance between two Boolean vectors α and β, denoted $d(\alpha,\beta)$, is the number of components in which they differ. If $d(\alpha,\beta) = 1$, then α and β are neighbors. A sequence of points $\alpha_1,\ldots,\alpha_k$ is called a path of length $k-1$ from α_1 to α_k if any two consecutive points in the sequence are neighbors. A shortest path between α and β is a path of length $d(\alpha,\beta)$. A true path is a path consisting only of true points of a Boolean function. Two true points are convexly connected if all the shortest paths connecting them are true. For an integer k, a Boolean function is called *k-convex* if and only if every pair of true points at distance at most k is convexly connected.

Convex functions have been studied in the context of LAD in [36]. The interest to convex functions is suggested by the fact that data points of the same type often exhibit certain compactness properties.

Decomposable functions. Let f be a Boolean function and $S_0, S_1, \dots, S_k$ be subsets of its variables. The function f is said to be *decomposable* with respect to the subsets $S_0, S_1, \dots, S_k$ if there exist Boolean functions $h_1, \dots, h_k$ and g such that

(i) h_i depends only on the variables in S_i, for each $i = 1, \dots, k$,
(ii) g depends on the variables in S_0 and on the binary values $h_i(S_i)$ for $i = 1, \dots, k$, i.e. $g: \{0,1\}^{|S_0|+k} \to \{0,1\}$,
(iii) $f = g(S_0, h_1(S_1), \dots, h_k(S_k))$.

To illustrate the usefulness of the notion of decomposability, let us assume that the components of the given binary vectors are associated with a list of food items, and that the positive and negative vectors represent observations on the days when a patient had or did not have a headache, while the sets S_1 and S_2, respectively, denote the (possibly overlapping) sets of food items containing proteins and carbohydrates; let S_0 represent the set of food items low in both proteins and carbohydrates. In order to test the hypothesis that insufficient variety in the types of proteins and in the types of carbohydrates during the day leads to headaches, we have to establish the existence or absence of a Boolean function $g(S_0, h_1(S_1), h_2(S_2))$ which separates correctly the positive and negative examples.

In [22], the authors study the computational complexity of the problem of determining if a partially defined Boolean function has a decomposable extension satisfying a given decomposition scheme and proved the following theorem.

Theorem 3.9. *Let $f = (T,F)$ be a partially defined Boolean function of n variables defined on $m = |T| + |F|$ points. Also, let $S_0, S_1, \dots, S_k$ be a family of subsets of its variables. If*

(a) $k = 1$ or
(b) $k = 2$ and $S_0 = \emptyset$ or
(c) $k = 2$ and $S_0 = O(\max(\log\log n, \log\log m))$,

then the problem of determining whether f admits an extension satisfying (i),(ii),(iii) can be solved in time polynomial in n and m. If

(d) $k = 2$ or
(b) $k = 3$ and $S_0 = \emptyset$,

then the problem of determining whether f admits an extension satisfying (i),(ii),(iii) is NP-complete.

This theorem says, in particular, that if a pair of sets S_0, S_1 of variables is specified in advance, then the problem of determining whether a partially defined Boolean function admits a respective decomposable extension is solvable in polynomial time. However, sometimes we want to know if the function has a pair of subsets S_0 and S_1 for which a decomposable extension exists. This problem was studied in [64] and was shown there to be NP-complete. More on finding decomposable extensions of a partially defined Boolean function can be found in [65, 66].

3.2.3.1 Bi-theories

Let $f = (T,F)$ be a partially defined Boolean function. To build a theory (i.e. to find an extension of f), LAD identifies a number of positive and negative patterns for f. Let us call the disjunction of positive patterns a *positive theory* and the disjunction of negative patterns a *negative theory*. An extension ϕ of f such that ϕ is a positive theory and $\overline{\phi}$ is a negative theory was called in [21] a *bi-theory*.

Example. Consider the pdBf of three variables defined by

$$T = \{(100),(111)\} \text{ and } F = \{(000),(001),(011)\}.$$

It is not difficult to verify by complete enumeration that the set of positive patterns consists of

$$x_1,\ x_1x_2,\ x_1\overline{x}_2,\ x_1x_3, x_1\overline{x}_3,\ x_1x_2x_3,\ x_1\overline{x}_2\overline{x}_3$$

and the set of negative patterns consists of

$$\overline{x}_1,\ \overline{x}_1x_2,\ \overline{x}_1\overline{x}_2,\ \overline{x}_1x_3,\ \overline{x}_1\overline{x}_3,\ \overline{x}_1x_2x_3,\ \overline{x}_1\overline{x}_2x_3,\ \overline{x}_2x_3,\ \overline{x}_1\overline{x}_2\overline{x}_3.$$

Therefore, $\phi = x_1$ is a bi-theory for (T,F), since x_1 is a positive pattern and $\overline{x}_1$ is a negative pattern. Also, $\phi = x_1x_2 \vee x_1\overline{x}_3$ is a bi-theory, since x_1x_2 and $x_1\overline{x}_3$ are positive patterns and $\overline{\phi} = \overline{x}_1 \vee \overline{x}_2x_3$ is a negative theory consisting of two negative patterns. It can be shown that there are no other bi-theories for this pdBf.

The notion of bi-theories was introduced in [21] with the objective to provide convincing justifications for classification of each individual point, rather than obtaining a high rate of correct classifications. In other words, the objective is the *a priori justification* of the rules rather than their *a posteriori performance*.

In [21], it was shown that every pdBf has bi-theory extensions. The simplest way of showing this is through the notion of decision trees.

A *decision tree* is a rooted directed graph in which the root has zero in-degree, every non-leaf vertex has exactly two outgoing arcs (left and right) and every leaf has zero out-degree. Each non-leaf vertex v is labeled by an index $j(v) \in \{1,2,\dots,n\}$ and the leaf vertices are labeled by either 0 or 1.

With each decision tree D one can associate a Boolean function $\phi_D : \{0,1\}^n \to \{0,1\}$ as follows. Let $x = (x_1,\dots,x_n)$ be a binary vector. Starting from the root, we move from vertex to vertex, always following the left arc out of v if $x_{j(v)} = 0$, and the right arc otherwise, and stop when we arrive at a leaf, in which case we say that x is classified into this leaf. The label of the leaf defines the value of $\phi_D(x)$.

Given a pdBf $f = (T,F)$, we say that a decision tree defines an extension of f if ϕ_D is an extension of f. Also, we say that a decision tree is *reasonable* for f if

- D defines an extension for f,
- for every leaf of D, at least one vector in $T \cup F$ is classified into the leaf.
- for every non-leaf vertex v, at least one vector from T is classified into a descendant of v, and at least one vector from F is classified into another descendant of v.

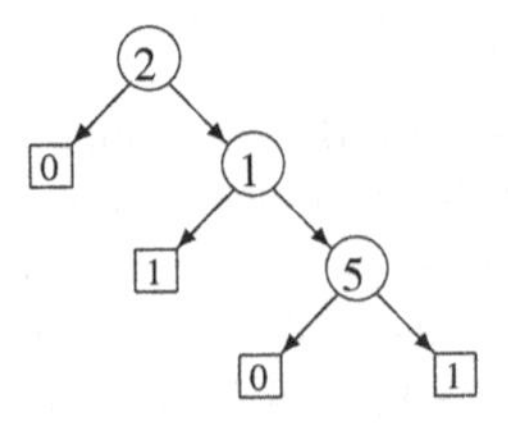

Fig. 3.1. An example of a decision tree

The importance of reasonable decision trees for partially defined Boolean functions is due to the following theorem proved in [21].

Theorem 3.10. *Let $f = (T,F)$ be a pdBf and D a reasonable decision tree for f. Then ϕ_D is a bi-theory of f.*

Many of the classical decision tree building methods yield reasonable trees. This fact together with the theorem above proves that every pdBf has bi-theory extensions. However, as was noted in [21], despite the strong relation between bi-theories and decision trees, the set of bi-theories for a pdBf is typically larger than the set of reasonable decision trees.

3.2.3.2 Extending Partially Defined Boolean Functions with Missing Bits

In practical situations the data may contain some errors. In [24, 25], the authors study the situation in which some bits in the given data set $T \cup F$ are missing, which may be due to erroneous data entry, the high cost of obtaining missing information, etc.

To model this situation, let us consider the set $M = \{0,1,*\}$ and interpret the asterisk components $*$ of a vector from M^n as missing (or uncertain) bits. Then a *partially defined Boolean function with messing bits* (*pBmb* for short) is a pair of disjoint subsets $T,F \in M^n$, where as before T is the set of true vectors and F is the set of false vectors.

Let S be the set of all missing bits of the pBmb (T,F). By assigning to each bit in S one of the two possible values 0 or 1, we obtain a partially defined Boolean function, which is called a *completion* of (T,F). Each assignment is a function γ: $S \to \{0,1\}$ and the number of such assignments is $2^{|S|}$. The completion of (T,F) obtained by assignment γ will be denoted (T^γ,F^γ).

For a completion (T^γ,F^γ) we denote by $\mathscr{E}(T^\gamma,F^\gamma)$ the set of all possible extensions of (T^γ,F^γ). In [24, 25], the authors follow a purely combinatorial approach, which leads to four different naturally arising notions of the extendibility of pBmbs. One of the most basic questions one has to answer in this setting is about the existence of a completion of a given pBmb, which has an extension belonging to a given class of Boolean functions $\mathscr{C}$.

Problem CE($\mathscr{C}$). Does there exist an assignment γ such that $\mathscr{E}(T^{\gamma},F^{\gamma})\cap\mathscr{C}\neq\emptyset$?

If the answer to the above question is "yes", then (T,F) is called consistent with respect to the class $\mathscr{C}$ and any function $f\in\mathscr{E}(T^{\gamma},F^{\gamma})\cap\mathscr{C}$ is called a *consistent extension* of (T,F) in $\mathscr{C}$.

The second naturally arising question is the existence of a function in the given class $\mathscr{C}$ which is an extension for all possible completions of the input pBmb.

Problem RE($\mathscr{C}$). Does there exist a function $f\in\mathscr{C}$ such that $f\in\mathscr{E}(T^{\gamma},F^{\gamma})$ for all possible assignments γ?

If yes, then f is called a *robust extension* of (T,F) with respect to $\mathscr{C}$. If a robust extension exists, it provides a logical explanation of (T,F) regardless of the interpretation of the missing bits.

An intriguing possibility, somewhere in between the previous two problems, is that a pBmb may not have a robust extension at all, despite the fact that all of its completions have extensions in the specified class.

Problem FC($\mathscr{C}$). Does there exist an extension $f_{\gamma}\in\mathscr{E}(T^{\gamma},F^{\gamma})\cap\mathscr{C}$ for every assignment γ?

If yes, then (T,F) is said to be *fully consistent* with respect to $\mathscr{C}$.

Let us note that the verification of affirmative answers for each of the previous three problems involves the description of a (or many) Boolean function $f\in\mathscr{C}$, which might be computationally expensive. Thinking of a computationally efficient derivation of robust extensions, we arrive at the notion of very robust extensions.

Let (T,F) be a partially defined Boolean function with messing bits, i.e. $T,F\subset\{0,1,*\}^{n}$. A term t is a called a *robust term* with respect to a vector $a\in T$, if $t(a^{\gamma})=1$ for all completions a^{γ} of a, and if $t(b^{\gamma})=0$ for all possible completions b^{γ} of all vectors $b\in F$. Let us note that, in particular, such a term cannot involve a component which is missing in a. Let us finally call a function $f\in\mathscr{C}$ a *very robust extension* of (T,F) in C if it can be represented as the disjunction $\vee_{a\in T}t_{a}$ in which each term t_{a} is robust with respect to the corresponding vector a.

Problem VR($\mathscr{C}$). Is there a very robust extension $f\in\mathscr{C}$ of (T,F)?

In [25], it was shown that Problem VR($\mathscr{C}$) is solvable in polynomial time for $\mathscr{C}$ being the class of all Boolean function, the class of positive functions and the class of k-DNF functions with fixed value of k. A DNF function is a Boolean function given via its DNF representation. It is a k-DNF if the degree of each term is at most k.

In contrast to the polynomial-time solvability of VR($\mathscr{C}$) in the class of k-DNF functions, Problem FC($\mathscr{C}$) in this class is co-NP-complete for each $k\geq 3$. This problem is also co-NP-complete in the class of threshold functions, while Problem RE($\mathscr{C}$) in this class is polynomial-time solvable. On the other hand, rather surprisingly, the three problems RE($\mathscr{C}$), FC($\mathscr{C}$) and VR($\mathscr{C}$) turned out to be equivalent for many general and frequently used classes of functions (two problems defined on the same set of input instances are said to be equivalent if for every input instance the problems yield the same answer; note that this notion is stronger than computational equivalence). In particular, in [25] the following result was proved:

Theorem 3.11. *Problems RE($\mathscr{C}$), FC($\mathscr{C}$) and VR($\mathscr{C}$) are all equivalent in the classes of all DNF-functions, 2-DNF functions and positive DNF-functions.*

3.3 Methodology

In this section, we turn to implementation of the basic ideas of LAD. Let us repeat that, speaking theoretically, the main objective of LAD is to find an extension of a partially defined Boolean function by means of revealing logical patterns hidden in the data. In practice, this general goal is frequently accompanied by a number of auxiliary problems and intermediate steps that have to be implemented to achieve the goal. We start by outlining these steps, updating along the way some terminology.

1. *Binarization*

In the introductory example presented in Section 3.1 the input data is given in the form of a partially defined Boolean function, i.e. the data consists of a number of binary vectors partitioned into two groups (of positive and negative points). In most real-life situations the input data is not necessarily binary and not necessarily numerical. The data comes as a collection of *observations* and this collection is frequently referred to as an *archive*. Each observation is an n-dimensional vector having as components the values of n *attributes*, also known as *features* or *variables*. For instance, the introductory example can be viewed as an archive of 7 observations, in which each observation is an 8-dimensional vector, and all 8 attributes are binary. Table 3.5 represents another example of an archive of 7 observations, but this time each observation is described by 4 attributes, where the first and the fourth attributes are numerical, while the second and the third ones are nominal. Each observation is accompanied by its "classification", i.e. by the indication of the particular class (positive or negative) this observation belongs to. To make these data amenable to the techniques of LAD, the archive has to be transformed into a binary format, i.e. into a partially defined Boolean function. This transformation is known as the binarization procedure and it is discussed in Section 3.3.1.

Table 3.5. A non-binary archive

x_1	x_2	x_3	x_4	class
1	green	yes	31	+
4	blue	no	29	+
2	blue	yes	20	+
4	red	no	22	+
3	red	yes	20	−
2	green	no	14	−
4	green	no	7	−

2. *Attribute selection*

Many real-life data sets contain exceedingly many attributes. For instance, the area of bioinformatics can frequently face problems involving tens of thousands entries. The binarization procedure can only increase this number. The presence of too many attributes in a data set, as well as the presence of confounding features, such as experimental noise, may lead – in the absence of adequate ways of selecting small, relevant subsets – to the failure of any attempt to extract knowledge from the data. The problem of selection of relevant sets of attributes is discussed in Section 3.3.2.

3. *Pattern generation*

After binarizing the data and selecting relevant attributes, LAD proceeds to the main procedure: generating patterns. The generation of patterns has always been the central issue in data analysis via LAD. In general, the number of patterns can be very large. Therefore, in practice, the generation is always restricted to patterns satisfying certain criteria. These criteria may include specification of the type of the patterns to be generated (e.g. prime or spanned) or specification of some pattern parameters (e.g. bounded degree). The choice of the criterion is problem-dependent and varies from data to data. In Section 3.3.3 we describe several algorithms for generating patterns satisfying various criteria.

4. *Model construction*

The pattern generation step produces a set of patterns called the *pandect* (sometimes by pandect they mean the collection of all patterns in the dataset). The number of patterns in the pandect may be too large to allow the effective utilization of all of them. Also, some of the patterns may occur only in rare cases, and they are redundant when more typical patterns are taken into account. All this leads to the problem of selecting a representative subset of patterns capable of providing classifications for the same set of points which can be classified by the pandect. The set of the selected patterns is called a *model*. LAD classifies observations on the basis of model's evaluation of them as follows. An observation (contained in the given set or not) satisfying the conditions of some of the positive (resp. negative) patterns in the model, and not satisfying the conditions of any of the negative (resp. positive) patterns in the model, is classified as positive (resp. negative). To classify an observation that satisfy both positive and negative patterns in the model, LAD constructs a *discriminant* (or *discriminating function*) that assigns relative weights to the patterns in the model. The problem of selecting patterns for a model and constructing a discriminant is discussed in Section 3.3.4.

5. *Validation and accuracy*

The last step of the LAD methodology is not specific to LAD and typical to all data mining techniques. It deals with validation of conclusions. We discuss it in Section 3.3.5.

3.3.1 Binarization

The logical analysis of data was originally developed for the analysis of datasets whose attributes take only binary values. Many real-life problems, however, use more complex data, especially *numerical* ones (temperature, weight, etc.), as well as *nominal* ones (color, shape, etc.). To make such problems amenable to the techniques of LAD, the problems have to be transformed into a binary format. A procedure for implementing this transformation was proposed in [23] and was called *binarization*.

The simplest non-binary attributes are the so-called "nominal" (or descriptive) ones. A typical nominal attribute is "shape", whose values can be "round", "triangular", "rectangular", etc. The binarization of a nominal attribute x can be done as follows. Let $\{v_1, \ldots, v_k\}$ be the set of all possible values of x that appear in the dataset. Obviously, this set is finite, since the number of observations in the dataset is finite. With each value v_i of x we associate a Boolean variable $\alpha(x, v_i)$ such that

$$\alpha(x, v_i) = \begin{cases} 1 & \text{if } x = v_i \\ 0 & \text{otherwise.} \end{cases}$$

The binarization of numerical attributes is different and is based on the notion of critical values, or *cutpoints*. Given a set of cutpoints for a numerical attribute x, the binarization of x consists in associating with each cutpoint t a Boolean variable x_t such that

$$x_t = \begin{cases} 1 & \text{if } x \geq t, \\ 0 & \text{if } x < t. \end{cases}$$

The new variable x_t is called the *indicator variable*.

Example. Consider Table 3.6 containing a set T of positive observations and a set F of negative observations and having three numerical attributes x, y, z.

Table 3.6. A numerical dataset

Attributes	x	y	z
T: positive observations	3.5	3.8	2.8
	2.6	1.6	5.2
	1.0	2.1	3.8
F: negative observations	3.5	1.6	3.8
	2.3	2.1	1.0

We introduce cutpoint 3.0 for the attribute x, cutpoint 2.0 for the attribute y and cutpoint 3.0 for the attribute z. This transforms the numerical attributes x, y, z into Boolean indicator variables $x_{3.0}$, $y_{2.0}$, $z_{3.0}$. The result of this binarization of Table 3.6 is given in Table 3.7.

Although exactly one cutpoint was introduced for each attribute in the above example, there is no reason to prohibit the introduction of more than one, or of no,

Table 3.7. A binarization of Table 3.6

Boolean variables	$x_{3.0}$	$y_{2.0}$	$z_{3.0}$
T: true points	1	1	0
	0	0	1
	0	1	1
F: false points	1	0	1
	0	1	0

cutpoint for some attributes. As another example, let us introduce cutpoint 2.0 for the attribute y and two cutpoints 5.0 and 1.5 for the attribute z and no cutpoints for the attribute x. The result of this binarization is shown in Table 3.8 (note that the first and the third vectors of T coincide).

Table 3.8. Another binarization of Table 3.6

Boolean variables	$y_{2.0}$	$z_{5.0}$	$z_{1.5}$
T: true points	1	0	1
	0	1	1
	1	0	1
F: false points	0	0	1
	1	0	0

In some cases, the choice of cutpoints is suggested by the nature of the attributes. For example, many medical parameters, such as body temperature, blood pressure, pulse rate, cholesterol level are termed "normal" or "abnormal" depending on whether they are inside/outside a certain interval or above/below a certain threshold. In those case where "critical" values of the attribute are unknown, a typical procedure for assigning cutpoints is as follows.

Let x be a numerical attribute. Since the number of observations in the dataset is finite, x can take only finitely many different values in the set. Let $v_1 < v_2 < \ldots < v_k$ be these values. Clearly, if we introduce two cutpoints between the consecutive values of x, then the corresponding Boolean variables will be identical. Therefore, it is sufficient to use at most one cutpoint between any two consecutive values of x. Also, cutpoints should be chosen in a way which allows to distinguish between positive and negative observations. In this respect, cutpoints below v_1 or above v_k are of no help. Therefore, one can be restricted to cutpoints of the form $\frac{1}{2}(v_{i-1}+v_i)$. A cutpoint $\frac{1}{2}(v_{i-1}+v_i)$ is called *essential* if there exist both a positive and a negative observation such that in one of them the value of x is v_{i-1}, while in the other $x = v_i$. Obviously, it suffices to use only essential cutpoints in the binarization procedure.

Except non-essential cutpoints, there may be further redundancy in the use of the remaining cutpoints. In [23], the authors consider the problem of finding a minimal set of cutpoints such that the corresponding binarized dataset is contradiction-free, i.e. allows to distinguish between positive and negative observations. This problem is phrased as a set-covering problem, which has an efficient greedy approximation algorithm, yielding a near-minimal number of cutpoints.

A modification of the greedy cutpoint selection procedure was proposed in [14]. It takes into account some kind of 'robustness' of the cutpoints chosen to binarize the data. The results obtained in [14] suggest that it is advantageous to minimize a combination of the number of cutpoints and their robustness and not simply the number of cutpoints.

One more approach to binarizing numerical data was proposed in [61]. It is based on a global study of the combinatorial property of a set of binary variable.

3.3.2 Attribute Selection

Whether a binary dataset is obtained by binarizing a numerical dataset or is generated naturally, it very likely to contain a number of redundant attributes. In order to prevent unsurmountable computational difficulties at the pattern generation stage, various techniques reducing the number of attributes have been developed in the literature. In this section, we first describe the standard LAD technique based on the notion of support sets of variables and then discuss some further ways to reduce the number of variables.

3.3.2.1 Support Sets of Variables

Let $f = (T,F)$ be a partially defined Boolean function with T being the set of its true points and F the set of its false points. A set S of variables is called a *support set* for f if f has an extension depending only on the variables from S. Equivalently, S is a support set if the projection of T on S is disjoint from the projection of F on S.

To better explain this notion, consider the false point 4 and the true point 5 of the partially defined function in Table 3.3. They differ only in the variables x_5 and x_6. Therefore, if both of these variables were omitted, the resulting points 4' and 5' would coincide, making the resulting pair (T',F') of projections not disjoint. This means that every support set of this function contains at least one of the two variable x_5 and x_6.

Clearly the set of all variables is a support set. However, a partially defined Boolean function may have support sets containing not all variables. The smallest support set selection problem consists in identifying a support set of minimum size. This is a problem of combinatorial optimization which can be solved in the following way.

Let us associate with every variable x_i a new binary variable y_i equal 1 if we include x_i in the support set, and equal 0 otherwise. Let further $\alpha = (\alpha_1 \dots \alpha_n)$ be a true point and $\beta = (\beta_1 \dots \beta_n)$ be a false point. Then at least one variable in which the two points differ must belong to any support set. This equivalent to requiring that

$$\sum_{i=1}^{n} (\alpha_i \oplus \beta_i) y_i \geq 1,$$

where $\oplus$ denotes the addition mod 2. Therefore, a minimum size support set of a partially defined Boolean function $f = (T,F)$ can be found by solving the following set covering problem:

$$\begin{aligned} \min & \sum_{i=1}^{n} y_i \\ s.t. & \sum_{i=1}^{n} (\alpha_i \oplus \beta_i) y_i \geq 1 \quad \text{for each pair } \alpha \in T \text{ and } \beta \in F \\ & y \in \{0,1\}^n \end{aligned} \tag{3.10}$$

For the function in Table 3.3, there are two solutions to this problem: $\{x_5, x_8\}$ and $\{x_6, x_7\}$. Both solutions can be found manually, since the size of the problem in this case is small: 8 variables and $3 \times 4 = 12$ constrains. In general, solving the set covering problem of this type is computationally difficult, i.e. NP-hard.

Several modifications of the basic set covering problem (3.10) have been proposed in [20, 30].

The gaol of the first modification is to assure that the true and the false points can be distinguished by more than one variable. In order to achieve this goal, the right-hand side of the inequality $\sum_{i=1}^{n} (\alpha_i \oplus \beta_i) y_i \geq 1$, should be replaced by a value higher than 1.

A further improvement of the quality of the support set can be achieved by equipping the variables in the objective function with coefficients (weights) reflecting the usefulness, or the quality of each variable. There are numerous ways of choosing values of these coefficients. One particular way, based on statistical measures, was proposed in [30]. More ways to measure the quality of variables are discussed in the next section.

3.3.2.2 Pattern-Based Feature Selection

One of the complicating factors in the extraction of a relevant subset of features (variables) is the fact that there is a marked difference between the relevance of individual features and that of subsets of features. In [5, 7], the authors propose an approach to feature selection that takes into account the "collective effect" of sets of features in distinguishing the positive and negative observations of a dataset. This approach consists of two major steps.

In the first step, a pool of features is selected based on several criteria for evaluating each one of the features in the dataset. This step can be viewed as a preprocessing procedure and it applies to the dataset before its binarization. Paper [7] describes five different criteria: separation measure, envelope eccentricity, system entropy, Pearson correlation and signal-to-noise correlation. For each of the criteria applied, only the top k ranked features retain for future consideration. The number k depends on the nature of the dataset and can be chosen experimentally. Both in [5] and in [7], $k = 50$.

In the second step, the dataset is binarized and a pandect of patterns is generated. The degree of participation of a variable in the patterns appearing in the pandect offers a valuable measure of its importance. In [5], the importance of a variable is defined as the ratio of the number of patterns containing the variable to the number of all generated patterns.

The pandect-based ranking of variables by their importance serves as a guiding principle for an iterative procedure of feature selection. Starting from the pool defined in Step 1, the procedure defines a new pool consisting of approximately half of the top ranked variables of the current pool, according to the pandect-based ranking.

3.3.3 Pattern Generation

In view of complete symmetry between positive patterns and negative patterns, in this section we restrict ourselves to the description of positive pattern generation.

In Section 3.2.2, we have distinguished several types of patterns. The two most frequently used are prime patterns and spanned patterns. In a sense, these two notions are opposite to each other. A pattern P is prime if it is inclusionwise maximal, i.e. any subcube properly containing P is not a pattern. A pattern P is spanned if it is an inclusionwise minimal pattern containing the points covered by P, i.e. any subcube properly contained in P covers fewer points than P.

Also, patterns are distinguished by three major parameters: degree, prevalence and homogeneity. We repeat that the degree of a pattern is the number of literals in it. In practice, patterns of small degree are always preferable because of its higher explanatory power. In other words, patterns of small degree are easier to interpret. Below we define the other two parameters.

The *prevalence* of a positive pattern is the ratio (sometimes expressed as the percentage) of the number of positive points covered by the pattern to the number of all positive points in the data set. The prevalence of a negative pattern is defined analogously. Obviously, patterns of high prevalence are more valuable.

In order to define the notion of homogeneity, we need to slightly relax the definition of a positive (negative) pattern. According to the original definition, a positive pattern is a subcube covering at least one positive and *no* negative point. In practice, finding such subcubes may result in patterns of very small prevalence. However, if we allow a subcube to cover a "few" negative points, the search may result in patterns with substantially higher prevalence. This observation justifies the following definition. The *homogeneity* of a positive pattern is the ratio (percentage) of the number of positive points covered by the pattern to the number of all points covered by it. The homogeneity of a negative pattern is defined analogously. Patterns of 100% homogeneity sometimes are called *pure* patterns.

In the next few sections we present several algorithms for the generation of pure patterns. In Section 3.3.3.1, we describe an algorithm for the generation of prime patterns of small degree proposed in [20]. In Section 3.3.3.2, we describe an algorithm for the generation of spanned patterns proposed in [9]. Also, in Section 3.3.3.3, we discuss two heuristics proposed in [19] for an approximate solution of the problem of finding maximum patterns. More information on pattern generation can be found in [6, 11].

3.3.3.1 Generation of Prime Patterns of Small Degree

Given a positive integer D, the algorithm outputs the set P_D of all positive prime patterns of degree at most D. A pseudocode of the procedure is presented as Algorithm 2 below. In what follows, we comment the main steps of the algorithm.

Algorithm 2. Algorithm for the enumeration of prime patterns up to a given degree

```
   Input: D the maximal degree D of patterns to be generated
   Output: P_D the set of prime patterns of degree up to D
 1 P_0 := ∅, C_0 := {∅};
 2 for d := 1 to D do
 3     P_d := P_{d-1}, C_d := ∅;
 4     for each term t in C_{d-1} do
 5         p := maximal index of literal in T;
 6         for s := p+1 to n do
 7             for ℓ_new ∈ {ℓ_s, ℓ̄_s} do
 8                 T := tℓ_new;
 9                 for i := 1 to d-1 do
10                     t' := T with i-th literal dropped;
11                     If t' ∉ C_{d-1}, then goto ◊;
12                 end
13                 If T covers a positive point but no negative point, then P_d := P_d ∪ {T};
14                 If T covers both a positive and a negative point, then C_d := C_d ∪ {T};
15                 ◊;
16             end
17         end
18     end
19 end
```

At each stage $d = 1,\ldots,D$ (loop 2–19 in the pseudocode), the algorithm produces two sets P_d and C_d. Each of the sets contains terms of degree d. The terms in P_d are positive prime patterns, while C_d contains so called "candidate" terms. A candidate term is a term covering at least one positive point and at least one negative point. Those terms of degree d that appear neither in P_d nor in C_d are irrelevant.

Initially, for $d = 0$ (line 1), the set P_d is empty, while the set C_d consists of the only term of degree 0, namely, the empty term.

For $d \geq 1$, the algorithm examines all candidate terms t of degree $d-1$ and by adding to them new literals generates terms T of degree d. To reduce the volume of computations, the algorithm examines the terms of C_{d-1} and generates new terms in a specific manner. To this end, the literals are ordered as follows: $x_1 \prec \overline{x}_1 \prec x_2 \prec \overline{x}_2 \prec \ldots$ and the terms of C_{d-1} are examined in the lexicographic order induced by this linear order. Then, a term $t \in C_{d-1}$ is extended to a term of degree d only by adding to it a literal which is larger (in this order) than any literal of t. To justify this modification, we let the indices of the literals in t be $i_1 < i_2 < \ldots < i_{d-1}$, and suppose that a term T is obtained from t by adding to it a literal of index $i < i_{d-1}$. Also, let t' be the term obtained from T by dropping the literal of index i_{d-1}. Both t and t' are terms of degree $d-1$, but t' is lexicographically smaller than t. Therefore, if t' also belongs to C_{d-1}, then T was already generated while examining t', in which case there is no need to generate it again while examining t. On the other hand, if t' does not belong to C_{d-1}, then T can be ignored according to following observation.

(1) If T contains a literal ℓ such that the term $t' := T - \ell$ does not belong to C_{d-1}, then T can be ignored. Indeed, in this case either $t' \in P_{d-1}$ (i.e. t' is a prime pattern of degree $d-1$) or t' is an irrelevant term of degree $d-1$.
   - If $t' \in P_{d-1}$, then T is neither prime (since it properly contains the prime pattern t') nor a candidate term (since it does not cover any negative point, as t' does not cover any negative point).
   - If t' is irrelevant, then T is obviously irrelevant too.

The above discussion allows us to be restricted to the generation of terms of degree d obtained from candidate terms of degree $d-1$ by adding to them literals with *large* indices (see line 5 and loop 6–17, where n denotes the number of variables). In line 8, a new term T of degree d is created, and in lines 9–12, the algorithm verifies if observation (1) is applicable to T. If so, the algorithm ignores T (by going to line 15)

If T covers a positive point but does not cover any negative point, then it is a positive pattern. Moreover, this pattern must be prime, since for every literal ℓ the term $T - \ell$ is not a positive pattern (as it belongs to C_{d-1}), in which case the algorithm adds T to the set P_d (line 13). If T covers both a positive and a negative point, then the algorithm adds it to the set C_d of candidate terms of degree d (line 14). If T does not cover any positive point, then the algorithm ignores T.

The efficient implementation of this algorithm requires some special data structure to store the sets C_d, which is described in [60].

3.3.3.2 Generation of Spanned Patterns

In this section, we describe an algorithm for the generation of all spanned patterns proposed in [9]. The algorithm works not only in the case of binary data, but also in the more general context of intervals described in Section 3.2.2.4.

The algorithm is based on a consensus-type method. Consensus-type methods enumerate all maximal objects of a certain collection by starting from a sufficiently large set of objects and systematically completing it by application of two simple operations: consensus adjunction and absorption.

For the generation of positive spanned patterns, the proposed method starts with a dataset of positive and negative observations and the collection C of positive patterns spanned by each individual positive observation. In the initial collection C, each pattern is the interval $[\alpha, \alpha]$, where α is a positive point.

The *consensus* of two patterns given by intervals $[\alpha,\beta]$ and $[\alpha',\beta']$ is the interval $[\alpha'',\beta'']$ with $\alpha_i'' = \min(\alpha_i,\alpha_i')$ and $\beta_i'' = \max(\beta_i,\beta_i')$. If the consensus of two positive patterns also is a positive pattern, the algorithm adds it to the collection.

The absorption operation in the proposed method coincides with the equality, i.e. a pattern P is said to absorb another pattern P' if simply $P = P'$.

In summary, the proposed algorithm for the generation of all positive spanned patterns starts with the collection C of patterns spanned by each individual positive observation and, as long as C contains a pair of patterns P and P' having consensus P'' not already in C, adds P'' to C. It was proved in [9] that this algorithm terminates and at termination the final list C contains all positive patterns spanned by subsets of positive observations.

In order to make the method more efficient, paper [9] proposes an accelerated version of the algorithm. The accelerated algorithm for the generation of all positive spanned patterns also starts with the collection C_0 of positive patterns spanned by each individual positive observation, but this collection remains unchanged during the execution of the algorithm. Along with this collection, the algorithm maintains the updated collection C, and pattern formation is restricted to pairs of patterns consisting of one pattern belonging to C_0 and one pattern belonging to C. This algorithm was called in [9] SPIC (spanned patterns via input consensus), and a pseudocode of the algorithm is presented below.

Algorithm SPIC

Let C_0 be the collection of patterns spanned by each individual positive observation.

Initiate $i := 0, C := C_0, W_0 := C_0$.
Repeat the following operations
  $W_{i+1} = \emptyset$
    For every pair of patterns $P_0 \in C_0$ and $P \in W_i$ having a consensus P' not contained in C, add P to W_{i+1} and to C.
  $i := i+1$
Until $W_i = \emptyset$.

It was proved in [9] that Algorithm SPIC generates all positive spanned patterns and the total running time of the algorithm is $O(\gamma n m m^+)$, where γ is the number of positive spanned patterns, n is the number of attributes, m is the number of observations in the dataset, and m^+ is the number of positive observations.

We repeat that the proposed method for the generation of spanned patterns is applicable to datasets which are not necessarily binary. Without any modification it can be applied to datasets whose attributes take values in the set $\{0,1,\ldots,k\}$ with $k \geq 1$. By means of a simple transformation, called discretization, every data set can be brought to this form (see e.g. [10]). We illustrate the application of Algorithm SPIC to a discrete (non-binary) dataset with the following simple example.

Example. Let the dataset consist of the four points (observations) $v_1 = (1,0,2)$, $v_2 = (1,1,1)$, $v_3 = (3,1,1)$, $v_4 = (2,0,2)$ with v_2 being negative and all other observations being positive.

- The input collection C_0 is $\{P_1 = [v_1, v_1], P_3 = [v_3, v_3], P_4 = [v_4, v_4]\}$. Initialize $C := C_0$.
- The consensus adjunction for the pair $P_1 \in C_0$ and $P_3 \in C$ results in the interval $P_{1,3} = [(1,0,1),(3,1,2)]$. This interval covers all four points, including the negative point v_2. Therefore, this is not a positive pattern and we do not include it in C.
- The consensus adjunction for the pair $P_1 \in C_0$ and $P_4 \in C$ results in the interval $P_{1,4} = [(1,0,2),(2,0,2)]$. This interval covers v_1 and v_4. Therefore, this is a positive pattern and we add it to C.
- The consensus adjunction for the pair $P_3 \in C_0$ and $P_4 \in C$ results in the interval $P_{3,4} = [(2,0,1),(3,1,2)]$. This interval covers v_3 and v_4. Therefore, this is a positive pattern and we add it to C.
- The consensus of any other pair from C_0 and C is either already in C or is not a positive pattern. The algorithm stops and outputs the family of all positive spanned patterns $C = \{P_1, P_3, P_4, P_{1,4}, P_{3,4}\}$.

3.3.3.3 Finding Maximum Patterns

Let us recall that for an observation α, an α-pattern is a pattern covering (i.e. containing) α. A maximum α-pattern is an α-pattern P with maximum coverage, i.e. with maximum number of positive points covered by P (if α is positive) or with maximum number of negative points covered by P (if α is negative). In Section 3.2.2.2, we described an integer programming and linear integer programming formulation for the problem of finding a maximum α-pattern. Both programs are computationally expensive, which may become an obstacle for solving the problem in case of very large datasets. To overcome this difficulty, in this section we describe two heuristics proposed in [19] that find an approximate solution.

In both heuristics we assume without loss of generality that α is a positive point. Also, in both heuristics we start with the characteristic term (minterm) of α, i.e. the term covering the point α only. This term can be described as $x_1^{\alpha_1} \ldots x_n^{\alpha_n}$, where

$$x_i^{\alpha_i} = \begin{cases} x_i & \text{if } \alpha_i = 1, \\ \overline{x}_i & \text{if } \alpha_i = 0. \end{cases}$$

Enlarging patterns to maximized prime patterns

Given a positive pattern P covering α, this heuristic transforms P into a positive *prime* pattern by successively removing literals from P. At each stage, the removal of a literal is considered to be advantageous if the resulting pattern is "closer" to the set of positive points not covered by it than to the set of negative points.

In order to specify the heuristic, let us define the *disagreement* between a point β and a pattern P to be the number of literals of P whose values at β are zero.

For instance, the disagreement between (001011) and $x_1x_3\bar{x}_4\bar{x}_6$ is 2, since x_1 and $\bar{x}_6$ take value 0 on this point, while x_3 and $\bar{x}_4$ take value 1. The disagreement between a set of points and a pattern is simply the sum of the disagreements between the pattern and every point in the set. Let us denote by $d_+(P)$ the disagreement between a positive pattern P and the set of positive points not covered by it. Similarly, $d_-(P)$ denotes the disagreement between P and the set of negative points. Computational experiments carried out in [19] suggest that the ratio $d_+(P)/d_-(P)$ provides a good criterion for choosing the literal to be removed at each step.

Enlarging patterns to maximized strong patterns

This heuristic extends the current positive pattern P covering α by choosing the next point to be included in $Cov(P)$, i.e. in the set of true points covered by P. For a non-empty subset S of true points, we denote by $[S]$ the *Hamming convex hull* of the points in S, i.e. the smallest subcube containing S. The heuristic criterion selects a true point β not covered by P such that $[Cov(P) \cup \{\beta\}]$ is a positive pattern with maximum number of literals.

3.3.4 Model Construction

The role of patterns in the LAD methodology is twofold. First of all, patterns provide a tool for classification of new observations, which is a typical task for many data mining techniques. On the other hand, patterns have the advantage of offering clear explanations of why a particular new observation is classified as positive or as negative. However, the consideration of an excessive number of patterns may diminish the explanatory power of LAD, which is due to the comprehensibility of each individual pattern rather than due to the comprehensiveness of a large collection of patterns (see [6]). Besides, too many patterns in the pandect may diminish the classification power of LAD, since in this situation many observations which are not in the dataset will be covered both by positive and by negative patterns. All this leads to the problem of selecting a representative subset of patterns capable of providing classifications for the same set of points which can be classified by the pandect. This collection is called a *model* and the problem of selecting patterns to be included in the model is discussed in Section 3.3.4.1.

In the ideal situation, every observation (contained in the set or not) is covered either by some positive and no negative patterns or by some negative and no positive patterns. In reality, many new observations will be covered both by positive and by negative patterns. To provide a classification for such observations LAD constructs a *discriminant* (or discriminating function) that assigns relative weights to the patterns in the model. This problem is discussed in Section 3.3.4.3.

3.3.4.1 Selection of Patterns

The role of this step is to select from the pandect of generated patterns a subset of patterns to be included in the model. This subset should, on the one hand, be of reasonable size, but, on the other hand, it should allow us to distinguish between the positive and the negative observations. In what follows, we describe the procedure for selecting positive patterns proposed in [20]. For the negative patterns the procedure is similar.

To each positive pattern P_k we assign a binary variable y_k with the convention that $y_k = 1$ if and only if P_k is included in the model. Also, for each positive observation point α_j let us define a binary vector $(a_{j1}, a_{j2}, \ldots, a_{jp})$, where p is the number of positive patterns and $a_{jk} = 1$ if and only if α_j is covered by the pattern P_k. Obviously, in order to distinguish α_j from the negative observations, at least one of the positive patterns covering it must be selected. In other words, the following inequality must be satisfied: $\sum_{k=1}^{p} a_{jk} y_k \geq 1$. In order to produce a small subset of patterns satisfying the requirements, we need to solve the following set covering problem:

$$\begin{aligned}
\min \; & \sum_{k=1}^{p} y_k \\
s.t. \; & \sum_{k=1}^{p} a_{jk} y_k \geq 1 \quad \text{for each positive observation } \alpha_j \\
& y_k \in \{0,1\} \quad k = 1, \ldots, p.
\end{aligned} \tag{3.11}$$

Let us observe that this formulation of the problem of pattern selection admits variations. For instance, we may require that each positive point be covered by several positive patterns, in which case we need to increase the right-hand side of the constraints. Also, in order to give preference to patterns possessing some special properties, such as low degree or high covering power, the objective function can be replaced by a weighted sum $\sum_{k=1}^{p} c_k y_k$ with appropriately chosen weights c_k. One more variation of the problem was studied [47] and is described in the next section.

3.3.4.2 Pattern Selection Taking into Account the Outliers and the Coverage of a Pattern

If an observation α is covered by a single pattern P from the generated pandect and the coverage of P is small, then α can be viewed as an outlier, in which case including P in the model is not justified. In addition, if a pattern covers only a few observations, including it in the model also is not justified (even if the observations covered by the pattern are not outliers) due to the low explanatory power of the pattern. To control this type of difficulties, the authors of [47] propose to enhance the set covering formulation (3.11) of the pattern selection problem by introducing a binary slack variable z_j for each observation α_j with the convention that α_j is an outlier if and only if $z_j = 1$. The requirement that an outlier does not need to be covered by a pattern can be achieved by modifying the right-hand side of the constraint in (3.11) as follows:

$$\sum_{k=1}^{p} a_{jk} y_k \geq 1 - z_j.$$

The objective function in (3.11) should be modified adequately to select a reasonable number of patterns. The modified objective function can be written as:

$$\min \sum_{k=1}^{p} y_k + C \sum_{j=1}^{n} z_j,$$

where n is the number of positive observations and C is a trade-off parameter between the number of selected patterns and outliers (i.e. trade-off between two conflicting goals: minimizing the number of selected patterns and minimizing the number of outliers). In [47], it was proved that if C satisfies $1/k < C \leq 1/(k-1)$, then the selected patterns cover at least k positive (negative) observations. Therefore, the formulation (3.11) can be modified as follows:

$$\begin{aligned} \min \ & \sum_{k=1}^{p} y_k + \frac{1}{k-0.5} \sum_{j=1}^{n} z_j & & \\ s.t. \ & \sum_{k=1}^{p} a_{jk} y_k \geq 1 - z_j & & \text{for each positive observation } \alpha_j \\ & y_k \in \{0,1\} & & k = 1,\ldots,p \\ & z_j \in \{0,1\} & & j = 1,\ldots,n, \end{aligned} \qquad (3.12)$$

where k is the desired coverage of the selected patterns. For $k = 1$ this formulation gives the same solution as (3.11).

3.3.4.3 Discriminant

The idea of the notion of discriminant is to emphasize the relative importance of patterns by assigning to them weights. To a positive pattern P_k we assign a positive weight w_k^+, and to a negative pattern N_l we assign a negative weight w_l^-. Then the discriminant is the following weighted sum:

$$\Delta(\alpha) = \sum_k w_k^+ P_k(\alpha) + \sum_l w_l^- N_l(\alpha),$$

where $P_k(\alpha)$ ($N_l(\alpha)$) is the value of P_k (N_l) at a point α (i.e. 1 or 0 depending on whether the point is covered or not by the pattern) . The weights of the patterns are chosen in such a way that a large positive (negative) value of the discriminant at a new observation point will be indicative of the positive (negative) character of that point.

If all weights have the same absolute value, then all patterns are equally important. On the other hand, the number q_k of observation points covered by a pattern P_k can be viewed as an indication of its relative importance, justifying the choice $|w_k| = q_k$. The relative importance of patterns can be emphasized even stronger by choosing $|w_k| = q_k^2$ or $|w_k| = q_k^3$ or $|w_k| = 2^{q_k}$. This approach can be generalized by choosing weights on the basis of appropriately defined distances from a pattern to the sets of positive and negative observations in the archive. Another reasonable point of view emphasizing the role of simple (i.e. short) patterns defines $|w_k| = 1/d_k$, where d_k is the degree of the pattern P_k.

In view of the possible disparity between the number of positive and of negative patterns, the weights may have to be normalized by a constant factor, assuring that

$$\sum_k w_k^+ = -\sum_l w_l^- = 1.$$

In the simplest case of equal weights, the normalized discriminant is calculated as

$$\Delta(\alpha) = \frac{\alpha_p}{p} - \frac{\alpha_n}{n},$$

where α_p and α_n are, respectively, the number of positive and the number of negative patterns covering α, while p and n are, respectively, the number of all positive and the number of all negative patterns in the model.

If $\Delta(\alpha)$ is positive, the observation α is classified as positive, and if $\Delta(\alpha)$ is negative, then α is classified as negative. LAD leaves unclassified any observation α for which $\Delta(\alpha) = 0$, since in this case either the model does not provide sufficient evidence, or the evidence it provides is contradictory. Computational experience with real-life problems has shown that the number of unclassified observations is usually small.

3.3.4.4 Large Margin Classifiers

The separation margin of a discriminant Δ is the difference between the smallest value that it takes over the positive points that are correctly classified and the largest value taken over the negative points that are correctly classifies. More formally, the separation margin is defined as

$$\min\{\Delta(\alpha) \ : \ \Delta(\alpha) > 0 \text{ and } \alpha \text{ is a positive point}\} -$$
$$\max\{\Delta(\alpha) \ : \ \Delta(\alpha) < 0 \text{ and } \alpha \text{ is a negative point}\}.$$

By maximizing the separation margin, one can expect a robust classification of unseen observations. The problem of finding an optimal discriminant was formulated in [18] as a linear program as follows:

$$\begin{array}{lll}
\max & p+n-C\sum_\alpha v_\alpha & \\
s.t. & \sum_k w_k^+ P_k(\alpha) + \sum_l w_l^- N_l(\alpha) + v_\alpha \geq p & \text{for each positive observation } \alpha \\
 & \sum_k w_k^+ P_k(\alpha) + \sum_l w_l^- N_l(\alpha) - v_\alpha \leq -n & \text{for each negative observation } \alpha \\
 & \sum_k w_k^+ = -\sum_l w_l^- = 1 & \\
 & p \geq 0, n \geq 0 & \\
 & w_k^+ \geq 0 & \forall k \\
 & w_l^- \leq 0 & \forall l \\
 & v_\alpha \geq 0 & \text{for each observation } \alpha,
\end{array} \tag{3.13}$$

where the sum in the objective function is taken over all observations in the data set, and

- P_k and N_l stand for positive and negative patterns, respectively,
- w_k^+ and w_l^- are the weights of the positive and negative patterns, respectively,
- p and n represent the positive and the negative part of the separation margin, respectively,
- v_α is the violation of the separating constraint corresponding to the observation α,
- C is a nonnegative penalization parameter that controls how much importance is given to the violations v_α.

Following [18], we refer to the above problem as RP. We purposely do not indicate the set of patterns over which RP is formulated. When applied to the set of patterns included in the model, a solution to this problem provides an optimal discriminant function for this set, i.e. finds the weights of the patterns that maximize the separation margin. However, that discriminant function may not be optimal with respect to the entire set of patterns in the generated pandect, or more generally, with respect to the set of all possible patterns. In order to verify global optimality, we need to make sure that there is no pattern that once added to the current set of patterns, allows for an improvement in the value of the objective function.

An approach to finding a global optimum was proposed in [18]. It starts with the initial set of patterns consisting of minterms (i.e. patterns each of which covers exactly one point) and solves RP to determine an optimal discriminant for this set. Then with the solution produced by RP the algorithm refers to a subproblem, called "pricing subproblem", which provides either a certificate of global optimality of the current discriminant function or a new positive or negative (or both) candidate pattern to be added to the current set of patterns aiming at the improvement of the global solution. Thus, this algorithm provides an alternative way to constructing a classifier which covers three steps of the traditional LAD approach: generating a pandect of patterns, selecting patterns from the generated pandect for the inclusion in the model, and constructing a discriminant. For the implementation details of the novel approach, we refer the reader to [18, 48].

3.3.5 Validation and Accuracy

When the original dataset is sufficiently large to allow the partition of the observations into a "training" and a "test set", the first one is used to derive a mathematical model and draw conclusions from it, while the second one is used to test the validity of the conclusions derived in this way. In case of small sets of observations, cross-validation techniques are frequently used for evaluating the quality of conclusions.

The most frequently used cross-validation technique is the usual k-folding method of statistics. This method consists in the random partitioning of the set of observations into k (approximately) equally sized subsets; one of these subsets is designated as the "test set", a model is built on the union of the remaining $k-1$ subsets (which form the "training set") and then tested on the k-th subset. This process is repeated k times by changing the subset taken as test set, and the average accuracy is then reported as a quality measure of the proposed method. A special case of

k-folding is the so-called "jackknifing", or "leave-one-out" technique, in which k is taken to be equal to the number of observations in the dataset, i.e. the test sets consist always of a single observation.

To calculate the quality of the model, let us denote by

a the percentage of positive observations that are correctly classified,
b the percentage of negative observations that are misclassified,
c the percentage of positive observations that are misclassified,
d the percentage of negative observations that are correctly classified,
e the percentage of positive observations that are unclassified,
f the percentage of negative observations that are unclassified.

Clearly, $a+c+e = 100\%$ and $b+d+f = 100\%$. The classification accuracy is defined as:

$$Q = \frac{a+d}{2} + \frac{e+f}{4}.$$

For theoretical analysis of the accuracy of LAD techniques, we refer the reader to [12, 13].

3.4 Applications

The LAD methodology outlined in the previous sections has found numerous applications across various fields, such as medicine, credit risk ratings, etc. A number of medical applications have been mentioned already in two survey papers [4, 42]. This includes cell growth prediction for polymeric biomaterial design [1], breast cancer prognosis by combinatorial analysis of gene expression data [2], logical analysis of diffuse large B-cell lymphomas [3], ovarian cancer detection by logical analysis of proteomic data [8, 67], coronary risk prediction by logical analysis of data [10, 54], and using LAD to differentiate entities of idiopathic interstitial pneumonias [28]. The LAD has been also applied to

- the early diagnosis of acute ischemic stroke in [68], where it has successfully identified 3 biomarkers that can detect ischemic stroke with an accuracy of 75%.
- identifying survival patterns for clear cell renal cell carcinoma [27]. The LAD survival score turned out to be more predictive of outcome than the Fuhrman classification system based on standard clinical parameters.
- revealing, for the first time, a correlation between the chemical structures of poly(β-amino esters) and their efficiency in transfecting DNA [37]. It was shown that detailed analysis of the rules provided by the LAD algorithm offered practical utility to a polymer chemist in the design of new biomaterials.
- detecting rogue components in the aviation industry [62], which substantially reduces time, costs and also increases the safety and overall performance of the operator.
- the analysis of survival data [53]. The performance of LAD when compared with survival decision trees improves the cross-validation accuracy by 18% for the gene-expression dataset.

- breast cancer diagnosis [52]. An important aspect of the analysis is that the authors found a sequence of closely related efficient rules, which can be readily used in a clinical setting because they are simple and have the same structure as the rules currently used in clinical diagnosis.
- modeling country risk ratings [39, 40]. The models include not only economic-financial but also political variables and provide excellent ratings even when applied to the following years data, or to the ratings of previously unrated countries.
- predicting secondary structure of proteins [17, 16], where LAD was applied to recognize which amino acids properties could be analyzed to deliver additional information, independent from protein homology, useful in determining the secondary structure of a protein.
- selecting short oligo probes for genotyping applications [50, 49]. Based on the general framework of logical analysis of data, the proposed probe design method selected a small number of oligo probes of length 7 or 8 nucleotides that perfectly classified all unseen testing sequences.
- identifying Chinese labor productivity patterns [38], which explain changes in productivity and lead to a decision support system aimed at increasing productivity of labor in China's provinces.

Among other applications, we can mention the use of LAD for differentiating chronic diffuse interstitial lung diseases [59], for characterizing relevant and irrelevant documents [26], for establishing morphologic code [58]. Some more of the applications of the LAD are described below in more detail.

3.4.1 Credit Risk Ratings

The progressively increasing importance of credit risk ratings is driven by the dramatic expansion of investment opportunities associated with the globalization of the world economies. Credit ratings published by agencies such as Moody's, Standard & Poor's, and Fitch, are considered important indicators for financial markets, providing critical information about the likelihood of future default.

LAD was applied in the area of credit risk ratings to two types of obligors: financial institutions and countries [39, 40, 43, 51]. In what follows we give a short overview of some results related to credit risk ratings of financial institutions obtained in [43].

The capability of evaluating the credit quality of banks has become extremely important in the last 30 years given the increase in the number of bank failures. The difficulty of accurately rating banks and other financial institutions is due to the fact that the rating migration volatility of banks is historically significantly higher than it is for corporations and countries, and that banks have higher default rates than corporations. Another distinguishing characteristic of the banking sector is the external support (i.e. from governments) that banks receive that other corporate sectors do not. Therefore, it is not surprising that the main rating agencies disagree much more often about the ratings given to banks than about those given to other sectors.

The Fitch individual bank credit rating system distinguishes five main ratings from A (very strong bank) to E (bank with very serious problems) and graduations among these ratings (A/B, B/C, C/D and D/E). A commonly used conversation of these ratings to a numerical scale is as follows: 9 corresponds to A, 8 to $A/B, \ldots$, 1 to E.

The dataset used in [43] consists of 800 banks rated by Fitch and operating in 70 different countries (247 in Western Europe, 51 in Eastern Europe, 198 in Canada and the USA, 45 in developing Latin-American counties, 47 in the Middle East, 6 in Oceania, 6 in Africa, 145 in developing Asian countries, and 55 in Hong-Kong, Japan, and Singapore).

The design of an objective and transparent bank rating system on the basis of the LAD methodology is guided by the properties of LAD as a classification system. To this end, the authors of [43] define a classification problem associated to the bank rating problem, construct an LAD model for it, and then define a bank rating system rooted in this model.

In this classification problem the banks rated by Fitch as A, A/B or B are defined as *positive observations* and the banks with ratings D, D/E or E are defined as *negative observations*. The original set of attributes consists of 24 numerical variables, such as loans, total earning assets, net interest revenue, overheads, equity, etc. However, five of the original variables turned out to be redundant, which was revealed by finding a *support set* of only 19 variables. All variables were binarized by means of cutpoints (the full list of which can be found in [51]). The model constructed in [43] is very parsimonious and consists of only 11 positive and 11 negative patterns, each of which is of degree at most 3. Three most important variables in the 22 patterns constituting the LAD model are credit risk rating of the country where the bank is located, the return on average total assets, and the return on average equity. The importance of the country risk rating variable, which appears in 18 of the 22 patterns, can be explained by the fact that credit rating agencies are reluctant to give an entity a better credit risk rating than of the country where it is located.

Mapping the numerical values of the LAD discriminant to the nine bank rating categories (A, A/B, ..., E) is accomplished using a non-linear optimization problem that partitions the interval of the discriminant values into nine sub-intervals associated to the nine rating categories. The LAD approach can also be used to derive models with higher granularity (i.e. more than non rating categories), which are used by banks to further differentiate their customers and to tailor their credit pricing policies accordingly.

This study showed that the LAD model cross-validates extremely well and is therefore highly generalized, and could be used by financial institutions to develop internal, Basel-compliant rating models.

3.4.2 Estimating Passenger Show Rates in the Airline Industry

A frequent practice in the airline industry is to overbook flights to make up for losses caused by absent passengers or to collect double fares if a prepaid passenger does not show up. This practice is not only used by the airline companies, but also in other industries, such as hotels or train transportation, where potential additional revenues can be generated when spaces are freed by absent customers.

Many methods exist for classification of booked passengers as shows or no-shows in order to improve the forecast of overbooking level. In [34], the LAD methodology was applied to study this problem. This study is based on a dataset called the PNR (Passenger Name Record). The information was collected for persons traveling from Vancouver to Calgary on flights commercialized by Air Canada during March 2009. This is equivalent to a population of 38 501 passengers to study.

Navigating through airlines datasets is an uneasy task; their volume and their structure make it difficult to understand for an outsider. Nevertheless, the authors of [34] extracted over 30 attributes for exploration. Although they did try some tests on this first set of attributes, it was rapidly noticed that not all attributes appear in the patterns. After filtering the attributes, this set was reduced to 16 attributes, such as day of the week, passenger's gender, frequent flyer program, origin of full itinerary, destination of full itinerary, point of sale, electronic ticket, round-trip booking, departure time, number in party (single, two passengers, etc), booking class, advance booking, etc.

Some of these attributes are either binary (frequent flyer program, electronic ticket, round-trip booking) or easy to binarize (gender). The remaining attributes were binarized by means of a cutpoint system. For instance, departure time was split by cutpoints into 8 intervals, origin of full itinerary into 10 zones according to geographic location. For the advance booking attribute, only one cutpoint was introduced (booked 60 days prior to departure), and for the day of the week attribute, 4 (instead of 6) cutpoints were used, since no change was observed in the show rates from Monday to Wednesday.

Many tests were made in order to explore the different combinations of homogeneity, prevalence and degree. As for the maximum pattern degree, it was found that patterns of degree 4 are optimal for the problem.

For simplicity sake, the authors of this study decided to keep all generated patterns in the model, i.e. they skipped the phase of pattern selection from the generated pandect. For the weights attached to each pattern in the discriminant, it was decided to use the relative prevalence.

When the model was constructed, it was tested on a new set of passengers, i.e. all travelers for the same market (Vancouver - Calgary) during April 2009. The results proved that LAD is superior to the commercial tool. When using a longer history for Air Canada's commercial tool, a full year of historical flights, the results were a little less impressive, but still clearly in favour of LAD. In addition, the LAD offers some explanation of the no-show causes when one looks into the content of the patterns and the usage of each attribute.

3.4.3 Improved Screening for Growth Hormone Deficiency Using Logical Analysis of Data

The growth hormone (GH) Research Society proposed criteria based on auxological parameters (height, difference between target and actual height, growth rate) to predict GH deficiency (GHD) followed by GH stimulation tests. The authors of [57] applied LAD with the objective to improve screening for GHD and so to reduce the need for stimulation tests. The study was done on the basis of a dataset consisting of 177 patients. The patients were split into 3 groups. Group 1 included 54 patients with GHD. Group 2 consisted of patients with short stature, who had a normal GH peak after pharmacological and sleep tests. Group 3 included 19 patients with transient GHD.

The patients in Group 1 were considered as positive observations and the patients in group 2 as negative observations. The authors of [57] identified 23 patterns in the dataset, of which 3 are negative and the remaining ones are positive. Classifiers were built using Group 1 and Group 2 only, because the diagnosis of these patients was clear, and were tested on Group 3.

One of the attributes in the dataset was plasma insulin-like growth factor (IGF) I concentration. The results showed that the IGF I value alone provides a more accurate diagnosis than the guidelines of the growth hormone Research Society. In conjunction with the growth rate, this factor creates a negative pattern which correctly predicted 52 of the 54 GHD cases. The authors concluded that IGF I, together with the growth rate, provides high quality diagnoses that are practical, simple and very accurate.

3.4.4 Using LAD for the Prediction of 1-Year Myocardial Infarction and Death

A common practice of predicting future cardiovascular risk is based on a number of traditional risk factors such as age, blood pressure, body mass index, etc. The authors of [29] hypothesized that identifying high- and low-risk patterns from a broad spectrum of hematologic phenotypic data may better predict future cardiovascular risk than traditional risk factors alone. Using logical analysis of data they performed a study which supports their hypothesis. In this study, a dataset of 7369 patients was first split into a derivation cohort (80% of patients) for model building and a validation cohort (20% of patients) for model testing.

In the derivation cohort, by means of LAD they identified 23 high-risk patterns and 24 low-risk patterns, all of degree 2. From a weighted sum of high- and low-risk patterns they derived a single calculated valued for an individual's overall 1-year risk of death or myocardial infarction. Within the validation cohort, this model demonstrated superior prognosis accuracy (78%) compared with traditional risk factors alone (67%).

3.4.5 Fault Detection and Diagnosis for Condition Based Maintenance Using the Logical Analysis of Data

Manufacturing firms face great pressure to reduce their production costs continuously. One of the main expenditure sources for these firms is maintenance costs. Presently, most maintenance decisions are based either on corrective maintenance, where actions are performed after system failure, on time-based preventive maintenance which sets a periodic interval to perform preventive maintenance regardless of the systems health. In contrast, condition-based maintenance (CBM) is a program that recommends actions based on the information collected through condition monitoring. In order to analyze the collected data, paper [70] applies Logical Analysis of Data. LAD is used for the identification of the machines state, in particular the state of potential failure called the faulty state, which begins by the transition from the state of normal (no fault) functioning. This state is critical to CBM because its identification allows the planning of maintenance actions, and the decrease of the risk of failure.

An LAD database contains measurements collected periodically of a number of machine state indicators. The measurements of the indicators collected at the same time, constitute an observation. By convention, the positive class is reserved to the faulty state while the negative class is the normal state.

The indicator values can be real numbers, discrete or continuous, nominal or numerical. The binarization process resulted in thirty nine binary attributes. It was decided to skip the feature selection step and keep the number of binary attributes as it is. The database consisted of 15 positive observations and 5 negative observations. From this data set, 10 of the positive observations and 3 of the negative observations are used for training. The remaining 5 positive observations and 2 negative observations are used to test the decision model.

This procedure was repeated 150 times, each time the combination of training and testing observations are changed. The selected maximum degree of the generated patterns was set to 1 for the first 50 trials, to 2 for the next 50 trials and to 3 for the last batch of 50 trials.

The patterns extracted from the data were given weights that represent the proportion of observations that are covered by these patterns.

The results obtained indicate classification accuracies equal to 95.86%, 93.43%, and 92.00%, for the maximum pattern degree equal to one, two and three respectively. In particular, this study resulted in the identification of two indicators, each of which alone correctly predicts the faulty state.

3.5 Conclusion

Logical Analysis of Data has made tremendous progress since the original publications by Peter L. Hammer [41, 32] and it continues to witness new theoretical developments and to find new applications. It also experiences new implementations, extensions, variations and combinations with other techniques. In particular,

- in [46], the patterns generated from the LAD are used as the input variables to the decision tree and k-nearest neighbor classification methods;
- in [63], LAD was used in combination with shadow clustering;
- [56] proposes extensions of LAD for growth hormone deficiency diagnoses;
- [55] introduces an LAD accelerator;

- [31] extends LAD techniques to finding three valued extensions of the input data and applies this methodology to a problem arising in molecular biology;
- [15] presents the LAD-CBM, the LAD-based software designed for diagnosis and prognosis in condition-based maintenance;
- [33] explores the idea of synthesizing artificial attributes from given ones. The artificial attributes carry increased amounts of information and allow a substantially more compact representation of information contained in a dataset. The application of this idea to biomedical datasets allowed the LAD to identify numerous new highly significant patterns of very low degree.

We conclude this survey by quoting several opinions of those who already used the LAD. These quotations speak for themselves.

"This technique avoids the drawbacks and limitations of the already-existing techniques, because of the transparency of its steps and the ease of interpretation of its results. Moreover, LAD is not based on any statistical assumptions and does not use any statistical theory" [70].

"The main advantage of LAD is that compared to classic statistical tools it can detect interactions between attributes (patterns), without any prior hypotheses, and patterns can be represented by simple rules. They are transparent objects, and can be easily understood by doctors and biologists. They are very useful biological research hypotheses and can be interpreted and further investigated by the experts" [53].

"The LAD is a classification method that has proven to be accurate, robust, flexible and adaptable. In addition, the LAD offers some explanation of the no-show causes when we look into the content of the patterns and the usage of each attribute" [34].

"LAD is capable of accomplishing in seconds something which takes days currently in the industry" [62].

"LAD provides the clinician with reliable information that can be used in daily practice" [59].

References

1. Abramson, S., Alexe, G., Hammer, P., Kohn, J.: A computational approach to predicting cell growth on polymeric biomaterials. J. Biomed. Mater. Res. Part A 73(1), 116–124 (2005)
2. Alexe, G., Alexe, S., Axelrod, D.E., Bonates, T.O., Lozina, I.I., Reiss, M., Hammer, P.L.: Breast cancer prognosis by combinatorial analysis of gene expression data. Breast Cancer Research 8, R41 (2006)
3. Alexe, G., Alexe, S., Axelrod, D.E., Hammer, P.L., Weissmann, D.: Logical analysis of diffuse large B-cell lymphomas. Artif. Intell. Med. 34, 235–267 (2005)
4. Alexe, G., Alexe, S., Bonates, T.O., Kogan, A.: Logical analysis of data — the vision of Peter L. Hammer. Annals of Mathematics and Artificial Intelligence 49, 265–312 (2007)
5. Alexe, G., Alexe, S., Hammer, P.L.: Pattern-based clustering and attribute analysis. Soft Comput. 10(5), 442–452 (2006)
6. Alexe, G., Alexe, S., Hammer, P.L., Kogan, A.: Comprehensive vs. comprehensible classifiers in logical analysis of data. Discrete Appl. Math. 156, 870–882 (2008)
7. Alexe, G., Alexe, S., Hammer, P.L., Vizvári, B.: Pattern-based feature selection in genomics and proteomics. Annals OR 148(1), 189–201 (2006)
8. Alexe, G., Alexe, S., Liotta, L.A., Petricoin, E., Reiss, M., Hammer, P.L.: Ovarian cancer detection by logical analysis of proteomic data. Proteomics 4(3), 766–783 (2004)
9. Alexe, G., Hammer, P.L.: Spanned patterns for the logical analysis of data. Discrete Appl. Math. 154, 1039–1049 (2006)
10. Alexe, S., Blackstone, E., Hammer, P.L., Ishwaran, H., Lauer, M.S., Snader, C.E.P.: Coronary risk prediction by logical analysis of data. Annals OR 119(1-4), 15–42 (2003)
11. Alexe, S., Hammer, P.L.: Accelerated algorithm for pattern detection in logical analysis of data. Discrete Appl. Math. 154, 1050–1063 (2006)
12. Anthony, M.: Accuracy of techniques for the logical analysis of data. Discrete Appl. Math. 96-97, 247–257 (1999)
13. Anthony, M.: Generalization error bounds for the logical analysis of data. Discrete Appl. Math. (to appear)
14. Anthony, M., Ratsaby, J.: Robust cutpoints in the logical analysis of numerical data. Discrete Appl. Math. 160, 355–364 (2012)
15. Bennane, A., Yacout, S.: LAD-CBM: new data processing tool for diagnosis and prognosis in condition-based maintenance. Journal of Intelligent Manufacturing (to appear)
16. Blazewicz, J., Hammer, P., Lukasiak, P.: Predicting secondary structures of proteins. IEEE Engineering in Medicine and Biology Magazine 24(3), 88–94 (2005)

17. Blazewicz, J., Hammer, P.L., Lukasiak, P.: Prediction of protein secondary structure using logical analysis of data algorithm. Computational Methods in Science and Technology 7(1), 7–25 (2001)
18. Bonates, T.O.: Large Margin Rule-Based Classifiers, pp. 1–12. John Wiley & Sons Inc. (2010)
19. Bonates, T.O., Hammer, P.L., Kogan, A.: Maximum patterns in datasets. Discrete Appl. Math. 156, 846–861 (2008)
20. Bores, E., Hammer, P., Ibaraki, T., Kogan, A., Mayoraz, E., Muchnik, I.: An implementation of logical analysis of data. IEEE Transactions on Knowledge and Data Engineering 12(2), 292–306 (2000)
21. Boros, E., Crama, Y., Hammer, P., Ibaraki, T., Kogan, A., Makino, K.: Logical analysis of data: classification with justification. Annals OR 188, 33–61 (2011)
22. Boros, E., Gurvich, V., Hammer, P.L., Ibaraki, T., Kogan, A.: Decomposability of partially defined boolean functions. Discrete Appl. Math. 62(1-3), 51–75 (1995)
23. Boros, E., Hammer, P., Ibaraki, T., Kogan, A.: Logical analysis of numerical data. Mathematical Programming 79, 163–190 (1997)
24. Boros, E., Ibaraki, T., Makino, K.: Logical analysis of binary data with missing bits. Artif. Intell. 107, 219–263 (1999)
25. Boros, E., Ibaraki, T., Makino, K.: Variations on extending partially defined boolean functions with missing bits. Inf. Comput. 180, 53–70 (2003)
26. Boros, E., Kantor, P.B., Neu, D.J.: Logical analysis of data in the TREC-9 filtering track. In: Proceedings of the Ninth Text REtrieval Conference (TREC-9), Maryland, USA, pp. 453–462 (2000)
27. Brannon, A.R., Reddy, A., Seiler, M., Arreola, A., Moore, D.T., Pruthi, R.S., Wallen, E.M., Nielsen, M.E., Liu, H., Nathanson, K.L., Ljungberg, B., Zhao, H., Brooks, J.D., Ganesan, S., Bhanot, G., Rathmell, W.K.: Molecular stratification of clear cell renal cell carcinoma by consensus clustering reveals distinct subtypes and survival patterns. Genes & Cancer 1(2), 152–163 (2010)
28. Brauner, M.W., Brauner, N., Hammer, P.L., Lozina, I., Valeyre, D.: Logical analysis of computed tomography data to differentiate entities of idiopathic interstitial pneumonias. In: Pardalos, P.M., Boginski, V.L., Vazacopoulos, A. (eds.) Data Mining in Biomedicine. Springer Optimization and Its Applications, vol. 7, pp. 193–208. Springer, US (2007)
29. Brennan, M.L., Reddy, A., Tang, W.H.W., Wu, Y., Brennan, D.M., Hsu, A., Mann, S.A., Hammer, P.L., Hazen, S.L.: Comprehensive peroxidase-based hematologic profiling for the prediction of 1-year myocardial infarction and death. Circulation 122(1), 70–79 (2010)
30. Bruni, R.: Reformulation of the support set selection problem in the logical analysis of data. Annals OR 150(1), 79–92 (2007)
31. Cepek, O., Kronus, D., Kucera, P.: Analysing dna microarray data using boolean techniques. Annals OR 188(1), 77–110 (2011)
32. Crama, Y., Hammer, P., Ibaraki, T.: Cause-effect relationships and partially defined boolean functions. Annals OR 16, 299–325 (1988)
33. Csizmadia, Z., Hammer, P.L., Vizvari, B.: Artificial attributes in analyzing biomedical databases. In: Pardalos, P.M., Hansen, P. (eds.) Data Mining and Mathematical Programming. CRM Proceedings and Lecture Notes, vol. 45, pp. 41–66. American Mathematical Soc. (2008)
34. Dupuis, C., Gamache, M., Page, J.F.: Logical analysis of data for estimating passenger show rates at Air Canada. Journal of Air Transport Management 18(1), 78–81 (2012)
35. Eckstein, J., Hammer, P.L., Liu, Y., Nediak, M., Simeone, B.: The maximum box problem and its application to data analysis. Comput. Optim. Appl. 23, 285–298 (2002)

36. Ekin, O., Hammer, P.L., Kogan, A.: Convexity and logical analysis of data. Theor. Comput. Sci. 244, 95–116 (2000)
37. Gubskaya, A.V., Bonates, T.O., Kholodovych, V., Hammer, P., Welsh, W.J., Langer, R., Kohn, J.: Logical analysis of data in structure-activity investigation of polymeric gene delivery. Macromolecular Theory and Simulations 20(4), 275–285 (2011)
38. Hammer, A., Hammer, P., Muchnik, I.: Logical analysis of chinese labor productivity patterns. Annals OR 87, 165–176 (1999)
39. Hammer, P., Kogan, A., Lejeune, M.: Modeling country risk ratings using partial orders. European Journal of Operational Research 175(2), 836–859 (2006)
40. Hammer, P., Kogan, A., Lejeune, M.: Reverse-engineering country risk ratings: a combinatorial non-recursive model. Annals OR 188, 185–213 (2011)
41. Hammer, P.L.: Partially defined boolean functions and cause-effect relationships. In: Lecture at the International Conference on Multi-Attrubute Decision Making Via OR-Based Expert Systems. University of Passau, Passau, Germany (1986)
42. Hammer, P.L., Bonates, T.O.: Logical analysis of data - An overview: From combinatorial optimization to medical applications. Annals OR 148(1), 203–225 (2006)
43. Hammer, P.L., Kogan, A., Lejeune, M.: A logical analysis of banks financial strength ratings. Technical Report TR-2010-9, The George Washington University (2010)
44. Hammer, P.L., Kogan, A., Simeone, B., Szedmák, S.: Pareto-optimal patterns in logical analysis of data. Discrete Appl. Math. 144, 79–102 (2004)
45. Hammer, P.L., Liu, Y., Simeone, B., Szedmák, S.: Saturated systems of homogeneous boxes and the logical analysis of numerical data. Discrete Appl. Math. 144, 103–109 (2004)
46. Han, J., Kim, N., Yum, B.J., Jeong, M.K.: Classification using the patterns generated from the logical analysis of data. In: Proceedings of the 10th Asia Pacific Industrial Engineering and Management Systems Conference, pp. 1562–1569 (2009)
47. Han, J., Kim, N., Yum, B.J., Jeong, M.K.: Pattern selection approaches for the logical analysis of data considering the outliers and the coverage of a pattern. Expert Systems with Applications 38(11), 13857–13862 (2011)
48. Hansen, P., Meyer, C.: A new column generation algorithm for logical analysis of data. Annals OR 188, 215–249 (2011)
49. Kim, K., Ryoo, H.: A lad-based method for selecting short oligo probes for genotyping applications. OR Spectrum 30, 249–268 (2008)
50. Kim, K., Ryoo, H.S.: Selecting genotyping oligo probes via logical analysis of data. In: Proceedings of the 20th conference of the Canadian Society for Computational Studies of Intelligence on Advances in Artificial Intelligence, CAI 2007, pp. 86–97. Springer, Heidelberg (2007)
51. Kogan, A., Lejeune, M.A.: Combinatorial methods for constructing credit risk ratings. In: Lee, C.F., Lee, A.C., Lee, J. (eds.) Handbook of Quantitative Finance and Risk Management, pp. 639–664. Springer, US (2010)
52. Kohli, R., Krishnamurti, R., Jedidi, K.: Subset-conjunctive rules for breast cancer diagnosis. Discrete Appl. Math. 154, 1100–1112 (2006)
53. Kronek, L.P., Reddy, A.: Logical analysis of survival data: prognostic survival models by detecting high-degree interactions in right-censored data. Bioinformatics 24, 248–253 (2008)
54. Lauer, M.S., Alexe, S., Pothier Snader, C.E., Blackstone, E.H., Ishwaran, H., Hammer, P.L.: Use of the logical analysis of data method for assessing long-term mortality risk after exercise electrocardiography. Circulation 106(6), 685–690 (2002)
55. Lejeune, M.A., Margot, F.: Optimization for simulation: Lad accelerator. Annals OR 188(1), 285–305 (2011)

56. Lemaire, P.: Extensions of logical analysis of data for growth hormone deficiency diagnoses. Annals OR 186(1), 199–211 (2011)
57. Lemaire, P., Brauner, N., Hammer, P., Trivin, C., Souberbielle, J.C., Brauner, R.: Improved screening for growth hormone deficiency using logical analysis data. Med. Sci. Monit. 15, 5–10 (2009)
58. Lupca, L., Chiorean, I., Neamtiu, L.: Use of lad in establishing morphologic code. In: Proceedings of the 2010 IEEE International Conference on Automation, Quality and Testing, Robotics (AQTR 2010), vol. 03, pp. 1–6. IEEE Computer Society, Washington, DC (2010)
59. Martin, S.G., Kronek, L.P., Valeyre, D., Brauner, N., Brillet, P.Y., Nunes, H., Brauner, M.W., Rety, F.: High-resolution computed tomography to differentiate chronic diffuse interstitial lung diseases with predominant ground-glass pattern using logical analysis of data. European Radiology 20(6), 1297–1310 (2010)
60. Mayoraz, E.: C++ tools for logical analysis of data. Report 1-95, Rutgers University, New Jersey, USA (1995)
61. Mayoraz, E.N., Moreira, M.: Combinatorial Approach for Data Binarization. In: Żytkow, J.M., Rauch, J. (eds.) PKDD 1999. LNCS (LNAI), vol. 1704, pp. 442–447. Springer, Heidelberg (1999)
62. Mortada, M.A., Carroll, T., Yacout, S., Lakis, A.: Rogue components: their effect and control using logical analysis of data. Journal of Intelligent Manufacturing (to appear)
63. Muselli, M., Ferrari, E.: Coupling logical analysis of data and shadow clustering for partially defined positive boolean function reconstruction. IEEE Trans. on Knowl. and Data Eng. 23, 37–50 (2011)
64. Ono, H., Makino, K., Ibaraki, T.: Logical analysis of data with decomposable structures. Theor. Comput. Sci. 289, 977–995 (2002)
65. Ono, H., Yagiura, M., Ibaraki, T.: An Index for the Data Size to Extract Decomposable Structures in LAD. In: Eades, P., Takaoka, T. (eds.) ISAAC 2001. LNCS, vol. 2223, pp. 279–290. Springer, Heidelberg (2001)
66. Ono, H., Yagiura, M., Ibaraki, T.: A decomposability index in logical analysis of data. Discrete Appl. Math. 142(1-3), 165–180 (2004)
67. Puszyński, K.: Parallel Implementation of Logical Analysis of Data (LAD) for Discriminatory Analysis of Protein Mass Spectrometry Data. In: Wyrzykowski, R., Dongarra, J., Meyer, N., Waśniewski, J. (eds.) PPAM 2005. LNCS, vol. 3911, pp. 1114–1121. Springer, Heidelberg (2006)
68. Reddy, A., Wang, H., Yu, H., Bonates, T., Gulabani, V., Azok, J., Hoehn, G., Hammer, P., Baird, A., Li, K.: Logical analysis of data (lad) model for the early diagnosis of acute ischemic stroke. BMC Medical Informatics and Decision Making 8(1), 30 (2008)
69. Ryoo, H.S., Jang, I.Y.: Milp approach to pattern generation in logical analysis of data. Discrete Appl. Math. 157, 749–761 (2009)
70. Yacout, S.: Fault detection and diagnosis for condition based maintenance using the logical analysis of data. In: 2010 40th International Conference on Computers and Industrial Engineering (CIE), pp. 1–6 (2010)

Peter L. Hammer (1936-2006)

In his life, Peter Hammer was a founder. He was the founder and Director of RUTCOR, Rutgers Center for Operations Research, which is now an internationally recognized center of excellence and an open institute, where seminars, workshops, graduate courses, and a constant flow of visitors create a buzzing and stimulating research environment.

He was the founder and editor-in-chief of several highly rated professional journals, including

- *Discrete Mathematics,*
- *Discrete Applied Mathematics,*

- *Discrete Optimization,*
- *Annals of Discrete Mathematics,*
- *Annals of Operations research,*
- *SIAM Monographs on Discrete Mathematics and Applications.*

He also was the founder of LAD, Logical Analysis of Data, to which part of this book is devoted. To emphasize the importance of data mining techniques, he launched the theme of "Discrete Mathematics and Data Mining" (which he called DM&DM), organized several workshops within the SIAM Data Mining Conferences, and edited special issues of *Discrete Applied Mathematics* devoted to these workshops.

After his untimely death in a tragic car accident on December 27, 2006, many tributes to his memory and many articles about his life appeared in the literature and in the Internet, see e.g. [2-7,9-12]. In the present note, we would like to share with the reader a few personal stories about this extraordinary man.

In the end of 1996, I (Vadim) was working on a generalization of a special method for computing the graph stability number invented by my Ph.D. advisor Vladimir Alekseev in the beginning of 80s. Browsing the Internet, I found, by pure chance, that the very same method was discovered under the name *struction* independently and approximately at the same time by another team of researchers: Christian Ebenegger, Peter Hammer, and Dominique de Werra [8]. I wrote an email to Peter and proposed to join our efforts in developing and exploring the generalization of struction method.[1] This was shortly before Christmas and I did not expect any reply until some weeks later. To my great surprise, a reply came in 15 minutes and this reply has changed my life as well as the life of our family.

In 2000, invited by Peter, we went to the USA for our first visit. Peter arranged for one of his Ph.D. students to meet us, which was very kind of him. What was much kinder is the fact that the refrigerator in our apartment was full of food, which was due to Peter again.

During the very first week of our visit, we have been invited to Peter's home for a dinner. Afterwards, we visited this hospitable home many times and had always enjoyed the unique atmosphere of the home and of the family. Peter's charming wife, Anca, was part of his life, his closest friend, who supported him in all possible respects. Typically, all conversations at the dinner table eventually ended up with "business matters", and if a guest wanted to apologize for this, Anca always said that this is what their life with Peter is.

Peter and Anca had always been very kind to our family, but one story was of particular impression. Our daughter, Yana, was a teenager at that time and lived in Russia, but she frequently visited us in the USA. One day she was returning

[1] This endeavor resulted in the paper [1] written jointly with Peter Hammer and Dominique de Werra. Christian Ebenegger passed away by that time and the special volume of *Discrete Applied Mathematics* where the paper appeared was partly devoted to Christian's memory.

to Russia by plane, but because of some technical problems the plane landed in a different airport far away from her home city. For a few hours we lost track of our daughter and of the plane. These were one of the most difficult hours of our life and Peter's help and support during these hours cannot be overestimated. He called the police, hospitals, airport managers, etc. Thanks God and Peter the situation was happily resolved.

Peter was insatiable in work. He worked everywhere, including his car while driving, and he expected the same attitude from his Ph.D. students. Often he finished his working day with the students late in the night,[2] and in the morning he already called asking for new results, ideas, etc.

Although Peter worked very hard, he always did it with joy and curiosity. He liked learning new things. For instance, in spite of the fact that he spoke several languages, including Russian, he always wanted to improve his skills. In particular, he bought an English-Russian dictionary and frequently spoke to us in Russian.

He also had a good sense of humor and frequently made fun or behaved in a funny way. For instance, he often called his students "victims". In his email to Irina he wrote: "Enjoy my absence, and work, work, work – as the great Lenin would have said if he would have been admitted to RUTCOR, but do not work more than 24-28 hours per day". In the Call for Papers for the first DM&DM meeting he expressed the hope that "the upcoming workshop will represent an important event in bringing DM closer to DM". In the Foreword to the first issue of *Discrete Applied Mathematics* (abbreviated DAM) he wished "to all of us who might have an interesting new result, which is Discrete, is Applied, and is Mathematics, a cordial DAM it!". When he chaired a session in a conference and wanted to let the speaker know that he is running out of time, Peter flashed the headlights of his electric wheelchair, or even honked.

We finish this note with a funny story, which shows, indirectly but very clearly, who was Peter Hammer. During the INFORMS conference in Florida in 2001 we met two colleagues who were shocked to hear that we give talks in the session chaired by Peter Hammer. "He must be over 100 years old", was their reply. Peter was not even 65 at that time, but he made already so many contributions to so many fields that people who did not know him personally thought of him as a legend rather than of a real human being. He is a legend indeed.

Vadim Lozin and Irina Lozina

Mathematics Institute, The University of Warwick

[2] The picture in the beginning of this note was taken by Irina in one of these "working nights". Later, it appeared in most articles devoted to Peter's life.

References

1. Alexe, G., Hammer, Peter L., Lozin, V. V., de Werra, D.: Struction revisited. Discrete Applied Mathematics **132**, 27–46 (2004)
2. Boros, E., Crama, Y., Simeone, B.: Peter Ladislaw Hammer. Discrete Optimization **4**, 257–259 (2007)
3. Berry, A., SanJuan, E., Pouzet, M., Golumbic, M. C.: Introduction to the special volume on knowledge discovery and discrete mathematics and a tribute to the memory of Peter L. Hammer. Annals of Mathematics and Artificial Intelligence **49**, 1–4 (2007)
4. Boros, E., Crama, Y., Simeone, B.: Peter Ladislaw Hammer. Discrete Math. **307**, 2153–2155 (2007)
5. Boros, E., Crama, Y., Simeone, B.: Peter Ladislaw Hammer. Discrete Appl. Math. **155**, 1345–1347 (2007)
6. Memoriam for Peter Ladislaw Hammer. Annals of Operations Research **149**, 1–2 (2007)
7. Boros, E., Crama, Y., Simeone, B.: Peter L. Hammer (1936–2006). 4OR **5**, 1–4 (2007)
8. Ebenegger, Ch., Hammer, P. L., de Werra, D.: Pseudo-Boolean functions and stability of graphs. *Algebraic and combinatorial methods in operations research*, North-Holland Math. Stud., Vol. 95, pp. 83–97, North-Holland, Amsterdam (1984)
9. Hartman, I. B., Naor, S., Penn, M., Rothblum, U.G.: Editorial [in memory of Professor Peter Hammer]. Discrete Appl. Math. **156**, 410–411 (2008)
10. Boros, E., Crama, Y., de Werra, D., Hansen, P., Maffray, F.: The mathematics of Peter L. Hammer (1936-2006): graphs, optimization, and Boolean models. Annals of Operations Research **188**, 1–18 (2011)
11. Peter Hammer. *Wikipedia*, `http://en.wikipedia.org/wiki/Peter_Hammer`
12. Peter Ladislaw Hammer (1936–2006). `http://rutcor.rutgers.edu/peterhammer.html`

Final Remarks

We have presented an overview of three approaches to the analysis of data: Test Theory (TT), Rough Sets (RS) and Logical Analysis of Data (LAD). We expect that this will stimulate research on further development of a variety of approaches, including methods based on combination of these and other approaches to data analysis, including machine learning, data mining, soft computing or granular computing approaches.

Scalability. Nowadays, one of the central issues in data analysis is scalability of the methods on huge data sets. We believe that there is a great potential in the discussed approaches for developing efficient scalable heuristics. Moreover, these approaches are very well predisposed for parallelization, for exploring of multithreading as well as for developing embedded hardware. For future cooperation between different teams, it will be also important to use platforms such as TunedIT (see `http://tunedit.org/`), ensuring the very important feature of reproducibility (of experiments) and making it possible for many teams to cooperate at a distance.

There are several areas of great importance for further development of intelligent systems that still need a lot of foundational studies.

Dynamic Networks of Data. For example, in data analysis, we are now moving from analysis of single data tables (or relational data bases) to dynamic networks of interacting sources of information. One of the examples of this growing research area is data stream analysis. There is a great need for deeper understanding of the nature of interactive computations and developing methods of data analysis in such interactive networks. Extending the existing methods from TT, RS and LAD to such networks is one of the great challenges. In building foundations of interactive computations on information granules in such networks all the discussed approaches will be of great importance.

Interactive Systems. In different areas such as nonmonotonic logics, physics or economy, it was recognized that moving from closed world (isolated system) to open interactive worlds creates a lot of challenges. Now, the data analysis area reached this stage too. The presented approaches TT, RS and LAD are very well predisposed to deal with these problems. One of the reasons for this is that all three approaches start from raw data over which other (hierarchical) levels of approximate reasoning can be built, contrary to many other approaches that attempt to build models of approximate reasoning by starting directly from higher levels and losing, in a sense, "contact" with raw data by making assumptions that are often inconsistent with real-life data. This feature already demonstrated that, e.g., RS can be successfully used in hybridization of experimental data with nonmonotonic reasoning in real-life projects such as control of unmanned helicopter (UAV) project.

Complex Adaptive Systems. Studying of dynamic networks of information sources will move us closer to problems of complex adaptive systems. Such systems are, in particular, autonomous and self-organized. One of the challenges is to develop methods for units (agents) in such systems for autonomous learning of protocols of cooperation, competition or coalition formation. Again, the approaches reported in the book can be of great importance in studies of these challenges. In particular, the experience in TT concerning fault detection can be used in developing strategies for self-healing in complex adaptive systems and all the approaches can be used in learning of complex behavioral patterns that are necessary, *e.g.*, for cooperation protocols or coalition formation.

Autonomous Systems. The current problems related to internet search engines or teams of robots cooperating with humans are moving us toward another exciting area of the research in which among nodes of dynamic information sources are human experts or users. Here, *e.g.*, rough set-based methods are important for approximation of information granules exchanged between different information sources. This helps agents to understand, at least to a degree, information received from other sources. In a broader sense, the three approaches discussed in this book can be also used in the growing research on semantic information and in moving the existing computers closer to natural language (e.g., see the idea of computing with words by Lotfi. A. Zadeh). The three approaches are also important for further development of methods in natural computing, in particular, for developing models of interactive computations based on perception.

Discrete vs. Continuous. Finally, there are issues related to the long-lasting discussion on relationships between the discrete and continuous approaches. These issues are nowadays not only philosophical but of great importance for applications. For example, one can search via data mining for relevant discrete or continuous patterns. We have already learned that for solving different problems, we also need methods for discovery and aggregation of both kind of patterns (*e.g.*, discrete and continuous models of processes). Here there is a room for fascinating research for all of the discussed approaches.

From Data Analysis to Reasoning about Data. Also, from the theoretical point of view, we should better understand relationships between reasoning based on Boolean functions and other forms of reasoning about data. This is moving us toward problems of reasoning about data that is strongly connected to the point of view of Leslie Valiant who mentioned the contradiction between logical nature of reasoning and the statistical nature of learning[1].

[1] http://people.seas.harvard.edu/~valiant/researchinterests.htm

Index